Chemistry

Revision Guide

Sue Orwin

Editor: Lawrie Ryan

OXFORD
UNIVERSITY PRESS

Contents

Introduction

Key points
- At the start of each topic are the important points that you must remember.
- Anything marked with the **H** icon is only relevant to those who are sitting the higher tier exams.

Synoptic link
Synoptic links show how the content of a topic links to other parts of the course. This will support you with the synoptic element of your assessment.

Study tip
Hints giving you advice on things you need to know and remember, and what to watch out for.

On the up
This feature suggests how you can work towards higher grades.

Using maths
This feature highlights and explains the key maths skills you need. There are also clear step-by-step worked examples.

This book has been written by subject experts to match the new 2016 specifications. It is packed full of features to help you prepare for your exams and achieve the very best that you can.

Key words are highlighted in the text. You can look them up in the glossary at the back of the book if you are not sure what they mean.

Many diagrams are as important for your understanding as the text, so make sure you revise them carefully.

Required practical
These practicals have important skills that you will need to be confident with for part of your assessment.

Anything in the higher tier spreads and boxes must be learnt by those sitting the higher tier exam. If you will be sitting foundation tier, you will not be assessed on this content.

Higher

In-text questions check your understanding as you work through each topic.

Summary questions
These questions will test you on what you have learnt throughout the whole chapter, helping you to work out what you have understood and where you need to go back and revise.

Practice questions
These questions are examples of the types of questions you may encounter in your exams, so you can get lots of practice during your course.

You can find brief answers to the summary questions and practice questions at the back of the book. More detailed exam-style feedback on practice questions is also available online.

Checklists
Checklists at the end of each chapter allow you to record your progress, so you can mark topics you have revised thoroughly and those you need to look at again.

Chapter checklist

Tick when you have:

reviewed it after your lesson	✓	☐	☐
revised once – some questions right	✓	✓	☐
revised twice – all questions right	✓	✓	✓

Move on to another topic when you have all three ticks

1 Atoms, bonding, and moles

Atoms are the chemical building blocks of our world. The periodic table organises these atoms and the elements they make into a structure that helps us to make sense of the physical world. Chemists have evidence that atoms themselves are made up of a nucleus with electrons surrounding it in energy levels. Theories of bonding explain how atoms are held together to make millions of different materials. Scientists use this knowledge of structure and bonding to engineer new materials with desirable properties.

Chemists use their calculations from quantitative analysis to determine the formulae of compounds and the equations for reactions. They also use quantitative methods to determine the purity of chemical samples and to monitor the yield from chemical reactions.

I already know...

I will revise...

I already know...	I will revise...
a simple model of the atom, representing atoms as hard, solid spheres of differing sizes and masses.	that atoms are made up of differing numbers of three different sub-atomic particles.
the differences between atoms, elements, and compounds.	how to explain the way atoms bond to each other in elements and in compounds.
how to use chemical symbols and formulae to represent elements and compounds.	how to explain the formula of elements and compounds, knowing the structure of the atoms and the type of bonding involved.
how to represent chemical reactions using formulae and using chemical equations.	how to carry out calculations using reacting masses to predict balanced symbol equations for reactions.
how patterns in reactions can be predicted with reference to the periodic table.	how to use atomic structure to explain patterns in reactivity in the periodic table.
the properties of metals and non-metals.	how to explain the difference between metals and non-metals in terms of their atomic structures and bonding.
the conservation of mass in chemical reactions.	how to carry out calculations using balanced symbol equations to predict the amounts of reactants and products in reactions.
how to use the particle model to describe changes of state.	how to describe changes of state and chemical reactions in terms of energy transfers.

1.1 Atoms

- There are about 100 different **elements** from which all substances are made. The **periodic table** is a list of the elements.
- Each element is made of one type of **atom**.
- Atoms are represented by chemical symbols. For example, Na for an atom of sodium, O for an atom of oxygen.
- The elements in the periodic table are arranged in columns, called groups. The elements in a **group** usually have similar properties.

1   What type of substances are shown in the periodic table?

- Atoms have a tiny **nucleus** surrounded by **electrons**.
- When elements react, their atoms join with atoms of other elements. **Compounds** are formed when two or more elements combine together.

2   What type of substance is sodium chloride, NaCl?

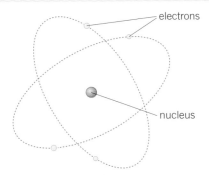

Each atom consists of a tiny nucleus surrounded by electrons

Key words: element, periodic table, atom, group, nucleus, electron, compound

Key points

- All substances are made of atoms.
- The periodic table lists all the chemical elements, with eight main groups each containing elements with similar chemical properties.
- Elements are made of only one type of atom.
- Compounds contain more than one element.
- An atom has a tiny nucleus at its centre, surrounded by electrons.

Synoptic link

Topic C2.1 gives more information on the periodic table of elements.

1.2 Chemical equations

- Chemical equations show the **reactants** (the substances you start with) and the **products** (the new substances made) in a reaction.
- **Word equations** give the names of the reactants and products only.

1   Write a word equation for the reaction between hydrogen and oxygen to produce water.

- In chemical reactions the atoms get rearranged. Symbol equations show the numbers and types of atoms in the reactants and products.
- Atoms are neither created nor destroyed in a chemical reaction. So the number and type of atoms remains the same before and after the reaction.
- When writing symbol equations you should always balance the equation. A **balanced symbol equation** has the same number of each type of atom on both sides of the equation. Use multipliers (numbers placed in front of formulae) to ensure symbol equations balance.
- You can include **state symbols** in balanced symbol equations. These are (s) for solids, (l) for liquids, (g) for gases, and (aq) for substances dissolved in water, called **aqueous solutions**.

Key points

- No new atoms are ever created or destroyed in a chemical reaction: the total mass of reactants = the total mass of products
- There is the same number of each type of atom on each side of a balanced symbol equation.
- You can include state symbols to give extra information in balanced symbol equations. The symbols are (s) for solids, (l) for liquids, (g) for gases, and (aq) for aqueous solutions.

Study tip

Make sure you NEVER change the formula of a reactant or product when balancing a symbol equation.

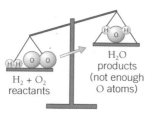
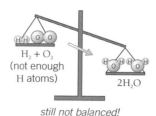
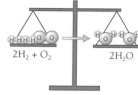

not balanced   *still not balanced!*   *balanced at last!*

Balancing an equation

On the up

You should be able to write balanced symbol equations including state symbols. For Higher Tier papers, you should also be able to write balanced half equations and ionic equations.

2   Balance this equation: $H_2 + Br_2 \rightarrow HBr$

3   H_2 is a gas. Which state symbol would you use to show a gas?

Law of conservation of mass

- The **Law of conservation of mass** states that no atoms are lost or made during a chemical reaction, so the mass of the products equals the mass of the reactants.

- In some reactions, the law seems to be broken because the mass appears to change. This is usually when gases are reactants or products. For example, if a product is a gas it may escape into the air. Sometimes substances increase in mass when heated in air. The extra mass is because oxygen gas from the air is a reactant.

Key words: reactants, products, word equation, balanced symbol equation, state symbols, aqueous solution, Law of conservation of mass

4   Which law states that the mass of the products formed in a reaction is equal to the mass of the reactants?

Student Book
pages 8–9   **C1**

1.3 Separating mixtures

- A mixture is made up of two or more substances (elements or compounds) that are not chemically combined together. The chemical properties of each substance in the mixture are unchanged.

Key points

- A mixture is made up of two or more substances that are not chemically combined together.
- Mixtures can be separated by physical means, such as filtration, crystallisation, and simple distillation.

- Mixtures are separated by physical processes. Physical processes do not involve chemical reactions, so no new substances are made. Physical processes include:

  ■ **Filtration**: separates substances insoluble in a solvent from those that are soluble in the solvent. For example, sand can be separated from salt solution using filtration.

  ■ Crystallisation: separates a soluble solid from a solvent, for example, salt (sodium chloride) from salt solution.

  ■ Distillation: separates a solvent from soluble solids dissolved in the solvent. For example, seawater is distilled to obtain usable water.

Synoptic links

To find out more about crystallisation and distillation of seawater, see Topics C5.5 and C14.2.

1   Name three methods of separating mixtures.

2   Which method would you use to separate sand from a mixture of sand and water?

Key word: filtration

1.4 Fractional distillation and paper chromatography

Key points

- Fractional distillation is an effective way of separating miscible liquids, using a fractionating column. The separation is possible because of the different boiling points of the liquids in the mixture.

- Paper chromatography separates mixtures of substances dissolved in a solvent as they move up a piece of chromatography paper. The different substances are separated because of their different solubilities in the solvent used.

Synoptic links

Topic C9.2 shows how fractional distillation finds use in the separation of crude oil.

You can see how to use paper chromatography to identify substances in Topic C12.2.

Key word: chromatography

- Fractional distillation is a way to separate mixtures of miscible liquids, for example, ethanol and water. Miscible liquids dissolve in each other, mixing completely. They do not form separate layers.

- The liquids have different boiling points; the liquid with the lowest boiling point is collected first.

- To aid separation you can add a fractionating column to the distillation apparatus.

- Fractional distillation is a way to separate ethanol from a fermented mixture in the alcoholic drinks industry.

1   In fractional distillation, which liquid is collected first?

- Paper **chromatography** is a way to separate substances from mixtures in solution. It works because some compounds are more soluble than others in the solvent.

- Paper chromatography is a way to separate food colourings.

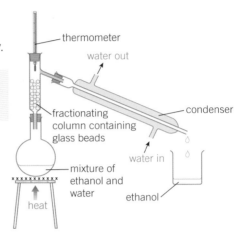

Fractional distillation of the miscible liquids, ethanol, and water

2   Name a method used to separate food dyes.

1.5 History of the atom

Key points

- Ideas about atoms have changed over time.

- New evidence has been gathered by scientists who have used their model of the atom to explain their observations and calculations.

- Key ideas were proposed successively by Dalton, Thomson, Rutherford, and Bohr.

- The ancient Greeks were the first to have ideas about atoms.

- In the early 1800s, Dalton linked his ideas to strong experimental evidence. Dalton suggested atoms were tiny, hard spheres. These atoms could not be divided or split.

- At the end of the 1800s, Thomson discovered a tiny negatively charged particle called the **electron**. Thomson proposed the 'plum pudding' model for the atom. The model suggested that negative electrons were embedded in a ball of positive charge. He imagined the electrons as the bits of plum in a plum pudding.

1   Who proposed the 'plum pudding' model?

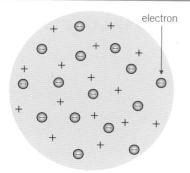

electron

sphere of positive charge

Thomson's 'plum pudding' model of the atom

Key words: electron, nucleus, proton, neutron

Study tip

Draw a timeline for theories relating to the structure of the atom.

- Ten years later, Geiger and Marsden were experimenting with alpha particles (dense positively charged particles). They fired alpha particles at a very thin piece of gold foil. A few alpha particles were repelled showing that there must be a tiny spot of positive charge in the centre of the atom. Rutherford then proposed the nuclear model.

- In the nuclear model, electrons orbit around a **nucleus** (centre of the atom). The nucleus contains positively charged **protons**.

- Bohr then revised the nuclear model. He suggested that the electrons were orbiting the nucleus in energy levels (or shells). The electrons were a set distance from the nucleus. Bohr's theoretical calculations matched the experimental observations.

- In 1932, Chadwick provided the experimental evidence that showed the existence of uncharged particles called **neutrons** in the nucleus.

On the up

To achieve the top grades, you should be able to justify why the model of the atom has changed over time.

Student Book pages 14–15 **C1**

1.6 Structure of the atom

Key points

- Atoms are made of protons, neutrons, and electrons.

- Protons have a relative charge of +1, and electrons have a relative charge of –1. Neutrons have no electric charge. They are neutral.

- The relative masses of a proton and neutron are both 1.

- Atoms contain an equal number of protons and electrons so carry no overall charge.

- Atomic number = number of protons (= number of electrons)

- Mass number = number of protons + neutrons

Key words: proton, neutron, electron, atomic number, mass number

- The nucleus at the centre of an atom contains two types of sub-atomic particle, called **protons** and **neutrons**. Protons have a positive charge and neutrons have no charge.

- **Electrons** are tiny negatively charged particles that move around the nucleus. An atom has no overall charge. That is because the number of protons is equal to the number of electrons and their relative charges are equal and opposite (proton +1 and electron –1).

- All atoms of an element contain the same number of protons. This number is called the **atomic number** (or proton number) of the element. Elements in the periodic table are arranged in order of their atomic number. The atomic number is also the number of electrons in an atom of the element.

- The **mass number** is the total number of particles in the nucleus of an atom, so it is the number of protons plus the number of neutrons.

Substitute numerical values into equations

Work out the number of each type of particle in an atom of fluorine from its atomic number of 9 and its mass number of 19.

Number of protons = atomic number = **9**

Number of electrons = number of protons = **9**

Number of neutrons = mass number – atomic number = 19 – 9 = **10**

1 How many protons, neutrons, and electrons are there in an atom of aluminium (atomic number 13, mass number 27)?

1.7 Ions, atoms, and isotopes

Key points

- Atoms that gain electrons form negative ions. If atoms lose electrons they form positive ions.

- You can represent the atomic number and mass number of an atom using the notation: $^{24}_{12}Mg$, where magnesium's atomic number is 12 and its mass number is 24.

- Isotopes are atoms of the same element with different numbers of neutrons. They have identical chemical properties, but their physical properties, such as density, can differ.

Synoptic link

For more information about the formation of ions from atoms, see Topic C3.2

Key words: ion, isotope

- An **ion** is a charged atom (or group of atoms).

- If an atom gains electrons, it becomes a negative ion. The ion has an overall negative charge because it has more electrons (–) than protons (+). For example, an oxygen atom gains two electrons to form a negative ion. The formula of the ion is written O^{2-}.

- If an atom loses electrons, it becomes a positive ion because it has more protons than electrons. For example, a lithium atom loses an electron to form a positive lithium ion, Li^+.

1 How does an atom form a positive ion?

- You can represent the atomic and mass number of an atom like this:

$$\text{mass number} \quad \begin{matrix} 12 \\ 6 \end{matrix} C \text{ (carbon)} \qquad \begin{matrix} 23 \\ 11 \end{matrix} Na \text{ (sodium)}$$
$$\text{atomic number}$$

- This shows that a sodium atom, $^{23}_{11}Na$, has an atomic number of 11 and a mass number of 23.

2 An atom of carbon is represented as $^{12}_{6}C$. What is the mass number of this carbon atom?

- You cannot see atoms because each individual atom is incredibly small. An atom is about a tenth of a billionth of a metre across (0.000 000 000 1 m, written in standard form as 1×10^{-10} m)

- Atoms of the same element always have the same number of protons. However, they can have different numbers of neutrons.

- Atoms of the same element with different numbers of neutrons are called **isotopes**.

- Isotopes of an element have different physical properties, but always have the same chemical properties.

3 Two isotopes of hydrogen are $^{1}_{1}H$ and $^{2}_{1}H$. State one difference between the two isotopes.

Standard form and nanometres

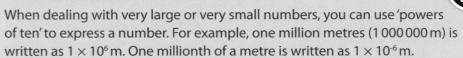

When dealing with very large or very small numbers, you can use 'powers of ten' to express a number. For example, one million metres (1 000 000 m) is written as 1×10^6 m. One millionth of a metre is written as 1×10^{-6} m.

Nanometres (nm) are used when describing the size of atoms, where 1 nm is equal to 1×10^{-9} m.

1.8 Electronic structures

Key points

- The electrons in an atom are arranged in energy levels or shells.
- The lowest energy level (1st shell) can hold up to two electrons and the next energy level (2nd shell) can hold up to eight electrons.
- After eight electrons occupy the 3rd shell, the 4th shell starts to fill.
- The number of electrons in the outermost shell of an element's atoms determines the way in which that element reacts.

Synoptic link

For more information on the reaction of elements and their electronic structures, see Chapter C2.

Key words: shells, electronic structure, noble gases

Study tip

Make a flow chart to describe how you would draw the electronic structure of the atoms for all of the first 20 elements when given their atomic number or their position in the periodic table (which tells you the number of electrons).

- Electrons are arranged around the nucleus in **shells**. Each shell represents a different energy level.

- Electrons occupy the lowest available energy levels (the shells closest to the nucleus).

- The first and lowest energy level (shell) can hold up to two electrons. The second energy level can hold up to eight electrons.

- You can show the arrangement of electrons in an atom by drawing diagrams or writing down the numbers of electrons in each energy level. After eight electrons occupy the third shell, the fourth shell starts to fill.

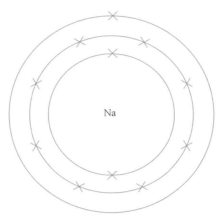

A simple way of representing the arrangement of electrons in the energy levels (shells) of a sodium atom

- The **electronic structure** shows the number of electrons in each energy level. You start recording the numbers of electrons at the lowest energy level. For example, sodium has 11 electrons, so sodium's electronic structure is 2,8,1

1   Aluminium has 13 electrons. Draw a diagram to show the arrangement of electrons in aluminium.

2   Silicon has 14 electrons. What is the electronic structure of silicon?

- Elements in the same group of the periodic table have the same number of electrons in their highest energy level. These electrons are called the outer electrons, because they are in the outermost shell. For example, all the Group 1 elements have one electron in their highest energy level (outer shell).

3   How many electrons are in the highest energy level of Group 2 elements?

- Because all the elements in the same group have the same number of electrons in their outer shell, they all react chemically in the same way.

- The elements in Group 0 of the periodic table are called the **noble gases**. They are very unreactive elements because their atoms have a very stable arrangement of electrons. Noble gases have eight electrons in their outer shell, except for helium, which only has two electrons in its outer shell.

1 **a** Name the elements in this list: Ca, CH_4, H_2, HCl, MgO, Ne, O_2, SO_2. [4 marks]

   b What general name is given to the other substances in the list? [1 mark]

2 The unbalanced equation for a reaction is:

$H_2 + Cl_2 \rightarrow HCl$

   a Balance the equation. [1 mark]

   b Write the word equation for the reaction. [1 mark]

3 Name a process you could use to separate ethanol and water. [1 mark]

4 What determines the order of the elements in the periodic table? [1 mark]

5 **a** Name the three types of sub-atomic particle. [3 marks]

   b State the relative mass and relative charge of each sub-atomic particle. [6 marks]

6 Describe how you could produce salt crystals from salt solution. [5 marks]

7 Draw a diagram to show the electronic structure of silicon (atomic number 14). [3 marks]

8 Why are magnesium and calcium in Group 2 of the periodic table? [1 mark]

9 Magnesium metal (Mg) reacts with dilute hydrochloric acid (HCl) to form a solution of magnesium chloride ($MgCl_2$) and hydrogen gas (H_2).
Write a balanced symbol equation, including state symbols, for this reaction. [3 marks]

10 The atomic radius of a silver atom is 1.44×10^{-10} m. Give the atomic radius of silver in nanometres. [1 mark]

11 There are three types of hydrogen atom: $_1^1H$, $_1^2H$, and $_1^3H$.

   a What name is used for these three types of atom? [1 mark]

   b Describe how the three types of atom are different. [1 mark]

12 Describe the differences between J.J. Thomson's 'plum pudding' model of the atom and Rutherford's nuclear model of the atom. [4 marks]

Chapter checklist

Tick when you have:

reviewed it after your lesson ✓ ☐ ☐

revised once – some questions right ✓ ✓ ☐

revised twice – all questions right ✓ ✓ ✓

Move on to another topic when you have all three ticks

1.1 Atoms ☐ ☐ ☐

1.2 Chemical equations ☐ ☐ ☐

1.3 Separating mixtures ☐ ☐ ☐

1.4 Fractional distillation and paper chromatography ☐ ☐ ☐

1.5 History of the atom ☐ ☐ ☐

1.6 Structure of the atom ☐ ☐ ☐

1.7 Ions, atoms, and isotopes ☐ ☐ ☐

1.8 Electronic structures ☐ ☐ ☐

2.1 Development of the periodic table

Key points

- The periodic table of elements developed as chemists tried to classify the elements. It arranges them in an order in which similar elements are grouped together.

- The periodic table is so named because of the regularly repeating patterns in the properties of elements.

- Mendeleev's periodic table left gaps for the unknown elements, which when discovered matched his predictions, and so his table was accepted by the scientific community.

- During the 19th century, scientists tried to find patterns in the behaviour of elements. At first, they arranged the elements in order of their atomic weight.

- In 1864, Newlands proposed his 'law of octaves'. He based the order on atomic weight and said that the properties of every eighth element were similar. However, new elements were being discovered that did not fit into his table. Other scientists did not accept his ideas and the elements in his octaves did not always have similar properties.

H 1	F 8	Cl 15	Co and Ni 22	Br 29	Pd 36	I 42	Pt and Ir 50
Li 2	Na 9	K 16	Cu 23	Rb 30	Ag 37	Cs 44	Os 51
Be 3	Mg 10	Ca 17	Zn 24	Sr 31	Cd 38	Ba and V 45	Hg 52
B 4	Al 11	Cr 19	Y 25	Ce and La 33	U 40	Ta 46	Tl 53
C 5	Si 12	Ti 18	In 26	Zr 32	Sn 39	W 47	Pb 54
N 6	P 13	Mn 20	As 27	Bi and Mo 34	Sb 41	Nb 48	Bi 55
O 7	S 14	Fe 21	Se 28	Rh and Ru 35	Te 43	Au 49	Th 56

Newland's table of octaves

1 How were the elements arranged in the early periodic tables?

- In 1869, Mendeleev placed the known elements in order of their atomic weights so that a pattern in their properties could be seen. In some places he changed the order of the elements so that elements with similar properties were in the same group. He left gaps for undiscovered elements.

- He used his periodic table to predict the properties of the undiscovered elements. When these elements were discovered they had similar properties to his predictions. Because of this, other scientists accepted his table.

- Later scientists discovered more about the structure of the atom. Knowledge of isotopes explained why some elements had heavier atomic weights than expected.

Dmitri Mendeleev on a Russian stamp issued in his honour in 1969. He is remembered as the father of the modern periodic table. Using the table, chemists could now make sense of the chemical elements

2 Why did Mendeleev leave gaps in his periodic table?

2.2 Electronic structures and the periodic table

Key points

- The atomic (proton) number of an element determines its position in the periodic table.
- The number of electrons in the outermost shell (highest energy level) of an atom determines its chemical properties.
- The group number in the periodic table equals the number of electrons in the outermost shell.
- The atoms of metals tend to lose electrons, whereas those of non-metals tend to gain electrons.
- The noble gases in Group 0 are unreactive because of their very stable electron arrangements.

Synoptic link

Topic C1.7 explains how atoms can gain or lose electrons to form ions.

- The elements are arranged in the periodic table in order of their atomic numbers (proton numbers). The elements are arranged in groups (vertical columns) with similar properties.
- The groups of elements have similar chemical properties because their atoms have the same number of electrons in their highest occupied energy level (outer shell).
- The group number equals the number of electrons in the outer shell of an atom.

1   Why do elements in a group have similar chemical properties?

- Metal elements are found on the left-hand side and centre of the periodic table. Metal elements tend to lose electrons and form positive ions.
- Non-metal elements are found on the right-hand side and towards the top of the periodic table. Non-metal elements tend to gain electrons and form negative ions.
- The atoms of noble gases in Group 0 are unreactive because of their very stable electron arrangements. They have eight electrons in the outermost shell (energy level) except for helium, which has just two electrons, but this complete first shell is also a very stable electronic structure.

2   Why are the noble gases so unreactive?

Non-metals   These elements generally have low melting and boiling points.

Group numbers

Relative atomic mass / Atomic (proton) number (1, H, 1)

Reactive metals   These metals react vigorously with other elements (like oxygen or chlorine), and with water. They are relatively soft – some of them can even be cut with a knife!

Elements 58–71 and 90–103 (all metals) have been omitted

Transition elements   This block contains the elements that most people probably think of when the word 'metal' is mentioned, like iron, copper, silver, and gold. These metals are not usually very reactive – some, like silver and gold, are very unreactive.

Noble gases   These (non-metal) elements are very unreactive, and it is very difficult to get them to combine with other elements.

The modern periodic table

Student Book
pages 26–27

C2

2.3 Group 1 – the alkali metals

Key points

- The elements in Group 1 of the periodic table are called the alkali metals.
- Their melting points and boiling points decrease going down the group.
- The metals all react with water to produce hydrogen and an alkaline solution containing the metal hydroxide.
- They form positive ions with a charge of 1+ in reactions to make ionic compounds. Their compounds are usually white or colourless crystals that dissolve in water producing colourless solutions.
- The reactivity of the alkali metals increases going down the group.

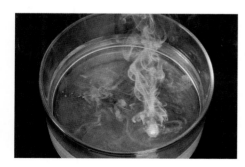

Potassium reacting with water

Key words: alkali metals, universal indicator

- The Group 1 elements are called the **alkali metals**. They are very reactive metals.
- Alkali metals are soft solids at room temperature with low melting and boiling points that decrease going down the group. They have low densities, so lithium, sodium, and potassium float on water.

1 Why are the alkali metals stored in oil?

- They react quickly with oxygen in the air. A layer of oxide forms on the surface of the metal. For example:

  sodium + oxygen → sodium oxide

  $4Na(s) + O_2(g) \rightarrow 2Na_2O(s)$

- They react with water to produce hydrogen gas and a metal hydroxide that is an alkali. For example:

  sodium + water → sodium hydroxide + hydrogen

  $2Na(s) + 2H_2O(l) \rightarrow 2NaOH(aq) + H_2(g)$

- The metal hydroxides are all soluble in water. They form a colourless solution with a high pH (**universal indicator** turns purple).

2 Why are the elements in Group 1 called 'alkali metals'?

- The alkali metal atoms all have one electron in their highest occupied energy level (outer shell). They lose this electron in reactions to form ionic compounds in which their ions have a single positive charge, for example, Na^+.
- They react with the halogens (Group 7) to form salts that are white or colourless crystals. For example:

  sodium + chlorine → sodium chloride

  $2Na(s) + Cl_2(g) \rightarrow 2NaCl(s)$

- Compounds of alkali metals dissolve in water, forming solutions that are usually colourless.
- Going down Group 1, the reactivity of the alkali metals increases.

7	Li
	lithium
3	
23	Na
	sodium
11	
39	K
	potassium
19	
85	Rb
	rubidium
37	
133	Cs
	caesium
55	
223	Fr
	francium
87	

The alkali metals (Group 1)

3 Name and give the formula of the compound formed when potassium reacts with bromine.

On the up

To achieve the top grades, you should be able to explain the trends in reactivity in the main groups in the periodic table in terms of electronic structure. See Topic C2.5.

2.4 Group 7 – the halogens

Key points

- The **halogens** all form ions with a single negative charge in their ionic compounds with metals.
- The halogens form covalent compounds by sharing electrons with other non-metals.
- A more reactive halogen can displace a less reactive halogen from a solution of one of its salts.
- The reactivity of the halogens decreases going down the group.

Synoptic links

For further information on the bonding between non-metal atoms in substances, see Topic C3.5, and for bonding between non-metal and metal ions in substances, see Topic C3.2..

Synoptic link

For further information on the state symbols (s), (l), (g), and (aq) that are used in equations, see Topic C1.2.

Key word: halogens

- The **halogens** are non-metallic elements in Group 7 of the periodic table.
- They exist as small molecules made up of pairs of atoms, for example, Cl_2. There is a single covalent bond between the halogen atoms in their molecules, for example, Cl–Cl.
- The halogens have low melting and boiling points that increase going down the group. At room temperature, fluorine is a pale yellow gas, chlorine is a green gas, bromine is a red-brown liquid, and iodine is a grey solid. Iodine easily vaporises to a violet gas.

Chlorine, bromine, and iodine

- They are all poor conductors of thermal energy and electricity.

1 Why do the halogens have low melting and boiling points?

- All of the halogens have seven electrons in their highest occupied energy level.
- The halogens form ionic compounds with metals in which the halide ions have a charge of 1–.
- The halogens also bond covalently with non-metals, forming molecules.
- The reactivity of the halogens decreases going down the group.
- A more reactive halogen is able to displace a less reactive halogen from an aqueous solution of a halide compound. For example:

chlorine + potassium iodide → potassium chloride + iodine

$$Cl_2(aq) + 2KI(aq) \rightarrow 2KCl(aq) + I_2(aq)$$

19	
	F
	fluorine
9	
35	
	Cl
	chlorine
17	
80	
	Br
	bromine
35	
127	
	I
	iodine
53	
210	
	At
	astatine
85	

The Group 7 elements, the halogens

2 How could you show that chlorine is more reactive than bromine?

On the up

You should be able to sketch graphs to show the trends in melting points, boiling points, and reactivity in Groups 1 and 7. You should also be able to use graphs to predict the properties of elements in the groups.

2.5 Explaining trends

Key points

- You can explain trends in reactivity as you go down a group in terms of the attraction between electrons in the outermost shell and the nucleus.

- This electrostatic attraction depends on:
  - the distance between the outermost electrons and the nucleus
  - the number of occupied inner shells (energy levels) of electrons, which provide a shielding effect
  - the size of the positive charge on the nucleus (called the nuclear charge).

- In deciding how easy it is for atoms to lose or gain electrons from their outermost shell, these three factors must be taken into account. The increased nuclear charge, due to extra protons in the nucleus, going down a group is outweighed by the other two factors.

- Therefore electrons are easier for the larger atoms to lose going down a group, and harder for them to gain going down a group.

Study tips

In Group 1, elements lose electrons when reacting, so reactivity **increases** going down the group.

However, in Group 7, elements gain electrons when reacting, so reactivity **decreases** as you go down the group.

Reactivity within groups

- Within a group the reactivity of the elements depends on the total number of electrons. Going down a group, there are more occupied energy levels and the atoms get larger. As the atoms get larger, the electrons in the highest occupied energy level (outer shell) are less strongly attracted by the nucleus.

- The inner shells of electrons 'screen' or 'shield' the outer electrons from the positive charge of the nucleus.

- When metals react they lose electrons, so the reactivity of metals in a group increases going down the group.

- When non-metals react they gain electrons, so the reactivity of non-metals decreases going down a group.

1 Why do metals get more reactive going down a group?

Explaining the trend in Group 1

- Reactivity increases going down Group 1 because the outer electron is less strongly attracted to the nucleus as the number of occupied energy levels increases and the atoms get larger.

- The attraction of the nucleus for the outer electrons is also reduced by the shielding effect of the inner shells.

this electron is easier to remove than the outer electron in a lithium atom

Li atom   Na atom

Sodium's outer electron is further from the nuclear charge and is shielded by more inner shells of electrons than lithium's outer electron

2 Why is lithium less reactive than sodium?

Explaining the trend in Group 7

- The reactivity of the halogens decreases going down Group 7 because the attraction of the outer electrons to the nucleus decreases as the number of occupied energy levels (shells) increases.

- More inner shells also shield the outer shell, reducing the attraction of the nucleus for an incoming electron.

2.6 The transition elements

- The **transition elements** are found in the central block of the periodic table between Groups 2 and 3.

- They are all metals and also called the transition metals.

45 Sc 21	48 Ti 22	51 V 23	52 Cr 24	55 Mn 25	56 Fe 26	59 Co 27	59 Ni 28	63.5 Cu 29	65 Zn 30
89 Y 39	91 Zr 40	93 Nb 41	96 Mo 42	98 Tc 43	101 Ru 44	103 Rh 45	106 Pd 46	108 Ag 47	112 Cd 48
	178 Hf 72	181 Ta 73	184 W 74	186 Re 75	190 Os 76	192 Ir 77	195 Pt 78	197 Au 79	201 Hg 80

The transition elements. The more well-known elements are circled

1 Where are transition metals found in the periodic table?

- Transition elements are good conductors of electricity and thermal energy.

- They are hard and strong with high densities.

- They have high melting points (except for mercury).

- The transition metals have very high melting points compared with Group 1 metals.

- Transition elements are harder, stronger, and much more dense than Group 1 metals.

- Transition elements are much less reactive than Group 1 metals. Transition metals only react slowly, or not at all, with oxygen and water at ordinary temperatures. Iron reacts much slower with chlorine gas than Group 1 metals do.

2 List the ways in which transition elements are different from the elements in Group 1.

- Transition elements form positive ions with various charges, for example, Fe^{2+} and Fe^{3+}.

- Compounds of transition metals are often brightly coloured.

- Many transition metals and their compounds are catalysts for chemical reactions.

Key points

- Compared with the alkali metals, transition elements have much higher melting points and densities. They are also stronger and harder, but are much less reactive.
- The transition elements do not react vigorously with oxygen or water.
- A transition element can form ions with different charges, in compounds that are often coloured.
- Transition elements and their compounds are important industrial catalysts.

Synoptic links

For more information on the bonding and structures of metals, see Topics C3.9 and C3.10.

Synoptic link

Catalysts are covered in Topic C8.5.

Key word: transition elements

Study tip

The charge on the metal ion is given in the name of many transition metal compounds, for example, copper(II) sulfate, $CuSO_4$, contains Cu^{2+} ions whereas copper(I) oxide, Cu_2O, contains Cu^+ ions.

1   How are elements arranged in the periodic table?   [2 marks]

2   What is the general name for the unreactive elements found in Group 0 of the periodic table?   [1 mark]

3   Why do groups of elements in the periodic table have similar properties?   [1 mark]

4   A small piece of lithium is added to a bowl of water.
    a   Write a word equation for the reaction of lithium with water.   [2 marks]
    b   Describe three things that you would see when the lithium is added to the water.   [3 marks]
    c   How could you show that an alkali is produced?   [2 marks]

5   Give one way in which the reaction of sodium with water is different from the reaction
    of lithium with water.   [1 mark]

6   Predict three physical and three chemical properties of the transition element cobalt, Co.   [6 marks]

7   What is the trend in melting points and boiling points going down Group 7?   [1 mark]

8   a   What is the formula of sodium bromide?   [1 mark]
    b   Describe its appearance and what happens when it is mixed with water.   [3 marks]

9   Some chlorine water is added to an aqueous solution of potassium bromide.
    a   Describe the colour change that you would see.   [2 marks]
    b   Write a word equation for the reaction that happens.   [2 marks]
    c   Write a balanced symbol equation for the reaction.   [2 marks]

10 How was Mendeleev's periodic table an improvement on Newlands table?   [2 marks]

11 Iron reacts with chlorine to produce iron(III) chloride.
    Write a balanced symbol equation for this reaction.   [2 marks]

12 Explain in terms of electronic structures:
    a   why sodium is more reactive than lithium   [4 marks]
    b   why fluorine is more reactive than chlorine.   [4 marks]

Chapter checklist

Tick when you have:

reviewed it after your lesson   ✓ ☐ ☐

revised once – some questions right   ✓ ✓ ☐

revised twice – all questions right   ✓ ✓ ✓

Move on to another topic when you have all three ticks

2.1 Development of the periodic table   ☐ ☐ ☐

2.2 Electronic structures and the periodic table   ☐ ☐ ☐

2.3 Group 1 – the alkali metals   ☐ ☐ ☐

2.4 Group 7 – the halogens   ☐ ☐ ☐

2.5 Explaining trends   ☐ ☐ ☐

2.6 The transition elements   ☐ ☐ ☐

3.1 States of matter

Key points

- The three states of matter are solids, liquids, and gases.
- The particle theory is used to explain the properties of solids, liquids, and gases.
- In melting and boiling, energy is transferred to the substance from the surroundings. In freezing and condensing, energy is transferred from the substance to the surroundings.
- **(H)** The simple particle model of solids, liquids, and gases is useful but has its limitations because the atoms, molecules, and ions that make up all substances are not solid spheres with no forces between them.

- The three states of matter are solids, liquids, and gases.
- All matter is made up of particles.
- The **particle theory** describes the movement and arrangement of particles.
- The particles are represented by small solid spheres.

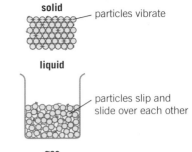

solid — particles vibrate

liquid — particles slip and slide over each other

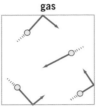

gas — particles move very quickly in all directions; as the particles bash against the walls of the container, they exert a force that causes pressure

The three states of matter

1 Name the three states of matter.

2 How are particles represented?

- Particles in a solid are packed closely together in a fixed arrangement. They vibrate constantly.

- Particles in a liquid are close together in a changing, random arrangement. They can move around.

- Particles in a gas are much further apart in a random arrangement. They move very quickly.

- A solid turns to a liquid at its melting point. As the temperature increases, the particles in a solid vibrate faster. At the melting point enough energy is transferred to the solid for the forces between particles to break. The particles break away from their fixed positions and start to move around. A liquid is formed.

- A liquid turns to a gas at its boiling point. As the temperature increases, the particles in the liquid move around faster. As more energy is transferred to the substance, some of its particles can escape from the surface of the liquid before the boiling point is reached. We say that the liquid is evaporating. At the boiling point, bubbles of gas form within the liquid and rise freely to the surface. A gas is formed.

- Substances with higher melting and boiling points have stronger forces between their particles.

- Each change of state is reversible. Gases condense to form liquids. Liquids freeze to form solids.

3 Which states contain particles in a random arrangement?

Key word: particle theory

4 Why do substances have different melting points?

Study tip

Close the book.

Draw labelled diagrams to show the arrangement and movement of particles in solids, liquids, and gases.

Limitations of the particle model

- The particle model is limited because it assumes particles are solid spheres with no forces between them.

- However, particles:
  - can be atoms, molecules or ions
  - vary in size
  - can contain many atoms
  - are not solid or spherical.

Higher

Student Book
pages 38–39
C3

3.2 Atoms into ions

Key points

- Elements react to form compounds by gaining or losing electrons, or by sharing electrons.

- The elements in Group 1 react with the elements in Group 7. As they react, atoms of Group 1 elements can each lose one electron to gain the stable electronic structure of a noble gas. This electron from the outer shell can be given to an atom from Group 7, which then also achieves the stable electronic structure of a noble gas.

- When two or more elements react together, compounds are formed. The atoms of elements join together by sharing electrons or by transferring electrons to achieve stable electronic structures. Atoms of the noble gases have stable electronic structures.

- When atoms of non-metallic elements join together by sharing electrons it is called **covalent bonding**.

  1 How can you tell that the compound H_2O has covalent bonds?

- When metallic elements react with non-metallic elements they produce ionic compounds. The metal atoms lose electrons to form positive ions. The atoms of non-metals gain electrons to form negative ions. The ions have the stable electronic structure of a noble gas. The oppositely charged ions attract each other in the ionic compound and this is called **ionic bonding**.

  2 Which of these compounds have ionic bonding?
     KBr   HCl H_2S Na_2O Cl_2O MgO

- Elements in Group 1 of the periodic table have atoms with one electron in their highest occupied energy level (outer shell). Sodium atoms, Na, (electronic structure 2,8,1), form sodium ions, Na^+ (electronic structure 2,8).

- Elements in Group 7 of the periodic table have atoms with seven electrons in their highest occupied energy level (outer shell). Chlorine atoms, Cl (2,8,7), form chloride ions, Cl^- (2,8,8).

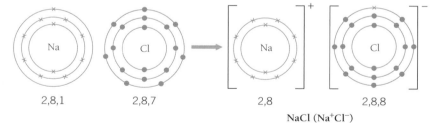

2,8,1   2,8,7   2,8   2,8,8

NaCl (Na^+Cl^-)

The formation of sodium chloride (NaCl) is an example of ion formation by transferring an electron

(2,8,1)   (2,8,7)   (2,8)   (2,8,8)

The dot and cross diagram to show electron transfer in sodium chloride, NaCl

Synoptic link

You can read about the formation of ions in Topic C1.7.

Key words: covalent bonding, ionic bonding, dot and cross diagram

- You can also use **dot and cross diagrams** to show the electrons in the outer shell of the atoms and ions involved in forming ionic bonds.

- The compound sodium chloride has equal numbers of sodium ions and chloride ions and so we write its formula as NaCl.

  3 Draw a dot and cross diagram to show how sodium atoms and chlorine atoms form ions.

3.3 Ionic bonding

- Ionic compounds are formed when metals react with non-metals.
- The oppositely charged ions formed are held together by strong forces of attraction, which act in all directions. This is called ionic bonding and a **giant structure** is formed. This can also be called a **giant lattice**.

Key points

- Ionic compounds are held together by strong forces of attraction between their oppositely charged ions. This is called ionic bonding.
- Besides the elements in Group 1 and Group 7, other elements that can form ionic compounds include those from Group 2 (forming 2+ ions) and Group 6 (forming 2– ions).

Key words: giant structure, giant lattice

1   How are ions held together in ionic bonding?

- Atoms from Group 1 form 1+ ions and atoms from Group 2 form 2+ ions.
- Atoms from Group 6 form 2– ions and atoms from Group 7 form 1– ions.

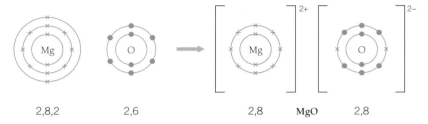

2,8,2   2,6   2,8   **MgO**   2,8

When magnesium oxide, MgO, is formed, the reacting magnesium atoms lose two electrons and the oxygen atoms gain two electrons

2   Calcium (Ca) is in Group 2. What is the symbol for a calcium ion?

On the up
Be able to draw dot and cross diagrams of unfamiliar ionic compounds.

3.4 Giant ionic structures

- Ionic compounds have giant structures in which many strong electrostatic forces, operating in all directions, hold the oppositely charged ions together. This means ionic compounds are solids at room temperature. It takes a lot of energy to overcome the ionic bonds to melt the solids. Therefore ionic compounds have high melting points and high boiling points.

Key points

- It takes a lot of energy to break the many strong ionic bonds, operating in all directions, that hold a giant ionic lattice together. So ionic compounds have high melting points. They are all solids at room temperature.
- Ionic compounds will conduct electricity when molten or dissolved in water. This is because their ions can then become mobile and can carry charge through the liquid.

1   Why do ionic solids have high melting points?

- However, when an ionic compound melts the ions are free to move. This allows them to carry electrical charge, so the liquids conduct electricity.
- Some ionic solids dissolve in water because water molecules can split up the lattice. The ions are free to move in the solutions and so they also conduct electricity.

strong electrostatic forces of attraction called ionic bonds

3D model of a giant ionic lattice

2 Why can ionic substances conduct electricity when molten or when dissolved in water?

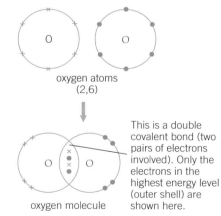

Student Book
pages 44–45

C3

3.5 Covalent bonding

- The atoms of non-metals need to gain electrons to achieve stable electronic structures. They can do this by sharing electrons with other atoms. Each shared pair of electrons strongly attracts the two atoms, forming a **covalent bond**. Substances that have atoms held together by covalent bonding are called molecules.

Key points

- Covalent bonds are formed when atoms of non-metals share pairs of electrons with each other.

- Each shared pair of electrons is a covalent bond.

- Many substances containing covalent bonds consist of simple molecules, but some have giant covalent structures.

Key word: covalent bond

hydrogen atoms (1)

hydrogen molecule

A shared pair of electrons gives both atoms a stable arrangement and forms a covalent bond.

Hydrogen gas is made up of H_2 molecules. Both H atoms in the H_2 molecule gain the stable electronic structure of the noble gas helium by sharing a pair of electrons

- A covalent bond acts only between the two atoms it bonds to each other, and so many covalently bonded substances consist of small molecules. Sometimes a molecule can contain double (or triple) covalent bonds, as in the oxygen molecule, O_2, shown below.

- Atoms of elements in Group 7 need to gain one electron and so form a single covalent bond. Atoms of elements in Group 6 need to gain two electrons and so form two covalent bonds. Atoms of elements in Group 5 can form three bonds and those in Group 4 can form four bonds.

oxygen atoms (2,6)

This is a double covalent bond (two pairs of electrons involved). Only the electrons in the highest energy level (outer shell) are shown here.

oxygen molecule

Oxygen gas is made up of O_2 molecules. Both O atoms in the O_2 molecule gain the stable electronic structure of the noble gas neon by sharing two pairs of electrons in a double covalent bond, often shown as O═O

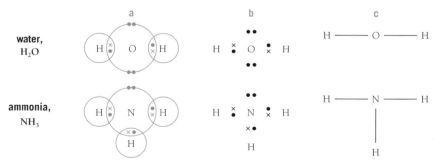

water, H_2O

ammonia, NH_3

*You can represent the bonding in a covalent compound by showing: **a** the highest energy levels (or outer shells), **b** the outer electrons in a dot and cross diagram, or **c** the number of covalent bonds*

1 Draw diagrams using symbols and lines to show the covalent bonds in: chlorine Cl_2, hydrogen chloride HCl, hydrogen sulfide H_2S, oxygen O_2, and carbon dioxide CO_2.

3.6 Structure of simple molecules

Key points

- Substances made up of simple molecules have low melting points and boiling points.

- The forces between simple molecules are weak. These weak intermolecular forces explain why substances made of simple molecules have low melting points and boiling points.

- Simple molecules have no overall charge, so they cannot carry electrical charge. Therefore, substances made of simple molecules do not conduct electricity.

- Models are used to help understand bonding but each model has its limitations in representing reality.

Using models

- You can use models to represent molecules. The models are useful but do have limitations.

- Most models do not show the actual shape of the molecule. All electrons are identical but dot and cross diagrams show electrons from different atoms differently. Electrons are shown in fixed positions, but they constantly move.

- In giant structures the models only represent a very tiny fraction of the millions of atoms (or ions) bonded together.

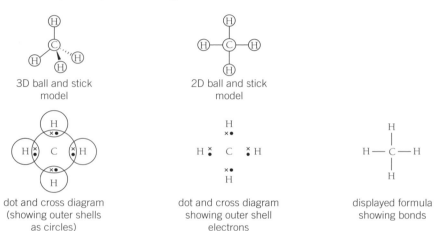

3D ball and stick model

2D ball and stick model

dot and cross diagram (showing outer shells as circles)

dot and cross diagram showing outer shell electrons

displayed formula showing bonds

The models used to help us understand the covalent bonding in a methane molecule

---

1 What are the limitations of dot and cross diagrams?

Intermolecular forces

- Substances made of small (simple) molecules have relatively low melting points and boiling points.

- Covalent bonds are very strong. This means covalent bonds hold the atoms within a molecule very tightly. However, each molecule is separate from neighbouring molecules and so simple molecules have little attraction for each other.

- The forces of attraction between molecules, called **intermolecular forces**, are weak. So not much energy is needed to overcome them. It is these forces that are overcome when a molecular substance melts or boils. This means that substances made of small molecules have low melting and boiling points.

---

2 Which forces are broken when a small molecule melts?

- Intermolecular forces increase with the size of the molecules; so larger molecules have higher melting and boiling points.

strong covalent bond

weak forces between molecules

*Covalent bonds and the weak forces between molecules in chlorine gas. It is the weak intermolecular forces that are overcome when substances made of simple molecules melt or boil. The covalent bonds are **not** broken*

Synoptic link

Topic C9.2 gives information on how the size of molecules affects their boiling points.

Key words: intermolecular forces, polymers

On the up

You should be able to explain which forces are broken when a substance containing simple molecules melts.

- **Polymers** have very large molecules. They are made up of many small molecules that covalently bond to each other to form long chains. For example, poly(ethene) is made up from thousands of ethene molecules, C_2H_4.

- Polymers are solids at room temperature because their intermolecular forces are relatively strong.

- Simple molecules do not conduct electricity because there is no overall charge on the simple molecules in a compound like sucrose.

3 Why does sucrose not conduct electricity?

poly(ethene)

where n is a large number

long chain of poly(ethene)

Two ways of representing the long polymer chains in poly(ethene)

Student Book
pages 48–49 **C3**

3.7 Giant covalent structures

Key points

- Some covalently bonded substances have giant structures. These substances have very high melting points and boiling points.

- The carbon atoms in diamond have a rigid giant covalent structure, making it a very hard substance.

- Graphite contains giant layers of covalently bonded carbon atoms. However, there are no covalent bonds between the layers. This means they can slide over each other, making graphite soft and slippery.

- Graphite can conduct electricity and thermal energy because of the delocalised electrons that can move along its layers.

Key words: giant covalent structures, delocalised electrons

- Some substances, such as diamond, form huge networks of atoms held together by strong covalent bonds in **giant covalent structures**. Every atom in the structure is joined to other atoms by strong covalent bonds. It takes an enormous amount of energy to break down the structure and so these substances have very high melting points.

- Diamond is a form of carbon with a giant covalent structure. Every carbon atom forms strong covalent bonds with four other carbon atoms. This makes diamond very hard, with a very high melting point.

- Diamond does not conduct electricity.

- As well as diamond, graphite and silicon dioxide (silica) have giant covalent structures.

1 Why does diamond have a very high melting point?

- In graphite, each carbon atom forms three strong covalent bonds with other carbon atoms. They form hexagonal rings, which are arranged in giant layers. There are no covalent bonds between the layers. Between the layers there are only weak intermolecular forces, so the layers can slide over each other quite easily. This makes graphite soft and slippery.

- In graphite, one electron from each carbon atom is delocalised, rather like electrons in a metal. These **delocalised electrons** allow graphite to conduct heat and electricity.

The structure of diamond

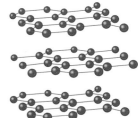

The giant structure of graphite

2 Give two similarities and two differences between diamond and graphite.

3.8 Fullerenes and graphene

Key points

- As well as diamond and graphite, carbon also exists as fullerenes. These can form large cage-like structures and tubes, based on hexagonal rings of carbon atoms.

- The fullerenes are finding uses transporting drugs to specific sites in the body, and also as catalysts and as reinforcement for composite materials.

- Graphene is a single layer of graphite and so is just one atom thick. Its properties, such as its excellent electrical conductivity, will help create new developments in the electronics industry in the future.

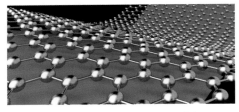

A computer-generated image of the structure of graphene

Key word: fullerenes

- **Fullerenes** have structures where carbon atoms join together to make large hollow shapes. The structure of fullerenes is based on hexagonal rings of carbon atoms. However, fullerenes can also have rings of five (pentagonal) or seven (heptagonal) carbon atoms.

- The first fullerene to be discovered was buckminsterfullerene (C_{60}), which contained 60 carbon atoms and had a spherical shape.

- Fullerenes have many important uses including drug delivery into the body, as lubricants, and as catalysts.

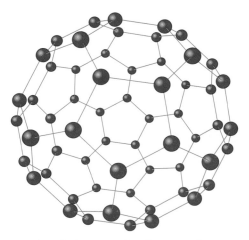

The first fullerene to be discovered was buckminsterfullerene (C_{60}), which contained 60 carbon atoms

1 Which element do all fullerenes contain?

- Cylindrical fullerenes called carbon nanotubes can also be produced. The nanotubes are very thin cylinders with a very high length to diameter ratio.

- Carbon nanotubes have very useful properties, such as high tensile strength that makes them useful to reinforce composite materials for making tennis rackets. They have delocalised electrons, giving high electrical conductivity so they are used in the electronics industry.

- Graphene is a single layer of graphite. It is a layer of hexagonal rings, one carbon atom thick.

- Graphene is an excellent conductor of electricity, has a very low density, and is incredibly strong for its mass. These properties make it very useful in electronics and composites.

2 Give one similarity and one difference between graphite and graphene.

3.9 Bonding in metals

Key points

- The atoms in metals are closely packed together and arranged in regular layers.

- You can think of metallic bonding as positively charged metal ions, which are held together by electrons from the outermost shell of each metal atom.

- The atoms in a metallic element are all the same size. They form giant structures in which layers of atoms are arranged in regular patterns.

1 How are the atoms arranged in a metal?

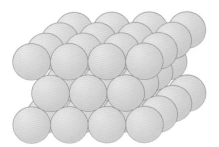

The close-packed arrangement of copper atoms in copper metal

- These delocalised electrons are free to move throughout the giant metallic lattice.

Key word: delocalised electrons

Study tip

Make a model of a metal structure using marbles or polystyrene balls.

- When metal atoms pack together, the electrons in the highest energy level (the outer electrons) delocalise and can move freely between atoms.

- This produces a lattice of positive ions in a 'sea' of moving electrons.

- The **delocalised electrons** strongly attract the positive ions and hold the giant structure together.

2   What forces hold metal atoms in place in their giant structures?

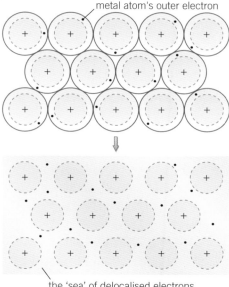

metal atom's outer electron

the 'sea' of delocalised electrons

A metal consists of positively charged metal ions surrounded by a 'sea' of delocalised electrons. This diagram shows us a model of metallic bonding

Student Book
pages 54–55   **C3**

3.10 Giant metallic structures

Key points

- Metals can be bent and shaped because the layers of atoms (or positively charged ions) in a giant metallic structure can slide over each other.

- Alloys are harder than pure metals because the regular layers in a pure metal are distorted by atoms of different sizes in an alloy.

- Delocalised electrons in metals enable electricity and thermal energy to be transferred through a metal easily.

Key word: alloys

- The atoms in a pure metal are arranged in closely packed layers in a giant metallic structure. The layers of atoms are able to slide easily over each other. This is why metals can be bent and shaped, but pure metals are too soft for many uses.

- **Alloys** are mixtures of metals or metals with other elements. The different sized atoms in the mixture distort the regular pattern of atoms in the layers of the metal structure. This makes it more difficult for the layers of atoms to slide over each other. As a result, alloys are harder than pure metals.

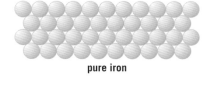

pure iron

iron alloy

Pure iron and an iron alloy

1   Why are alloys often more useful than pure metals?

- Metals have high melting and boiling points. In the giant structure the electrostatic forces of attraction between positive metal ions and delocalised electrons extend in all directions. Therefore a lot of energy is required to break down the lattice and separate the metal ions to form a liquid.

- Metals are good conductors of thermal energy and electricity. The electrical charge and thermal energy are transferred quickly through the giant metallic structure by the free-moving delocalised electrons.

2   Why are metals good conductors of electricity?

C3 **3.11 Nanoparticles**

Key points

- Nanoscience is the study of small particles that are between 1 and 100 nanometres in size.

- Nanoparticles may have properties different from those for the same materials in bulk. This arises because nanoparticles have a high surface area to volume ratio, with a high percentage of their atoms exposed at their surface.

- Nanoparticles may result in smaller quantities of materials, such as catalysts, being needed for industrial processes.

- **Nanoscience** deals with particles that measure between 1 and 100 nm and are just a few hundred atoms in size.

- 1 nanometre (1 nm) $= 1 \times 10^{-9}$ metres ($= 0.000\ 000\ 001$ m or a billionth of a metre).

1   What is a nanoparticle?

- Scientists refer to particles in air, for example, pollutants, pollen, or dust, as particulate matter (PM). The differently-sized particles have different names. Depending on their size, particles can be called nanoparticles, fine particles, or coarse particles.

- As fine and coarse particles are bigger than nanoparticles, scientists also use micrometres (μm, where 1 μm $= 1 \times 10^{-6}$ m) to describe their size.

- Nanoparticles have diameters between 1 nm and 100 nm (1×10^{-9} m to 1×10^{-7} m). Fine particles ($PM_{2.5}$) have a diameter of between 0.1 μm (1×10^{-7} m or 100 nm) and 2.5 μm (2.5×10^{-6} m or 2500 nm). Coarse particles (PM_{10}) have a diameter of between 2.5 μm (2.5×10^{-6} m or 2500 nm) and 10 μm (1×10^{-5} m).

- Coarse particles are often called dust.

- Nanoparticles can be 100 times smaller than even the finest dust.

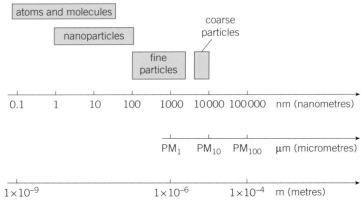

The size of different types of particles

2   A nanoparticle is 0.000 000 005 m in size. Write 0.000 000 005 m in standard form.

3   How many nanometres is 0.000 000 005 m?

Key word: nanoscience

Study tip

Write a tweet to explain what a nanoparticle is.

- Materials may behave very differently as nanoparticles. Nanoparticles' high surface area to volume ratio makes them much more reactive than materials with normal particle sizes. This is because nanoparticles have a very high proportion of their atoms or molecules at the surface of the particle.

- The use of nanoparticles instead of traditional bulk materials should mean that smaller quantities are needed.

3.12 Applications of nanoparticles

Nanoparticles at work

- A major new area of science is research into nanoparticulate materials.

- In cosmetics, nanoparticles are used in deodorants and in face creams where they are absorbed deeper into the skin. The nanoparticles used in sun creams are more effective at blocking the Sun's rays than normal-sized particles.

- In medicine, carbon nanocages are used to deliver drugs in the body. In wound dressings, coatings of silver nanoparticles are used to help to protect against bacteria.

- In computers, nanowires give vastly improved memory capacities and speeds.

- The large surface area of nanoparticles makes them very effective as catalysts.

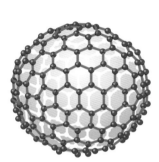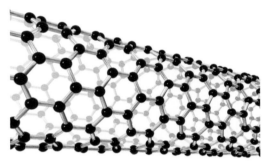

Nanocages can carry drugs inside them and nanotubes can reinforce materials

1 Give two uses of nanotechnology.

Possible risks

- Although many scientists think there are few health risks, there are some worries about nanoparticles.

- Other scientists believe that the health hazard increases as the diameter of the particles decreases. Breathing in tiny particles could damage the lungs. Nanoparticles could enter the bloodstream this way, or from their use in cosmetics, with unpredictable effects on our cells.

- Nanoparticles could enter the environment, for example, after washing clothing impregnated with silver nanoparticles. This could affect aquatic life by accumulating in organisms over time.

- More research, including long-term studies, needs to be done to find out about the effects of nanoparticles on health and the environment.

2 State two possible risks of using nanoparticles.

1 What type of structure is diamond? [1 mark]

2 What is an alloy? [1 mark]

3 **a** Which electrons in an atom are involved in bonding? [1 mark]

    b What happens to the electrons in atoms when ionic bonds are formed? [2 marks]

    c What happens to the electrons in atoms when covalent bonds are formed? [1 mark]

4 How are the atoms in metals arranged? [2 marks]

5 **a** Why do the elements in Group 1 form ions with a single positive charge? [3 marks]

    b Why do the elements in Group 7 form ions with a single negative charge? [3 marks]

6 Draw a dot and cross diagram to show the bonding in ammonia, NH_3. [2 marks]

7 Graphene is a single layer of graphite.
Suggest how graphene conducts electricity. [3 marks]

8 Describe the arrangement and movement of particles in a gas. [4 marks]

9 Why are nanoparticles very effective catalysts? [2 marks]

10 Explain why sodium chloride has a high melting point. [3 marks]

11 Explain why substances made up of simple molecules are gases at room temperature. [3 marks]

12 **a** A nanoparticle has a diameter of 70 nm. What is its diameter in m?
Give your answer in standard form. [1 mark]

    b A dust particle has a diameter of 2×10^{-7} m. What is its diameter in nm? [1 mark]

Chapter checklist

Tick when you have:

reviewed it after your lesson	✓		
revised once – some questions right	✓	✓	
revised twice – all questions right	✓	✓	✓

Move on to another topic when you have all three ticks

3.1 States of matter

3.2 Atoms into ions

3.3 Ionic bonding

3.4 Giant ionic structures

3.5 Covalent bonding

3.6 Structure of simple molecules

3.7 Giant covalent structures

3.8 Fullerenes and graphene

3.9 Bonding in metals

3.10 Giant metallic structures

3.11 Nanoparticles

3.12 Applications of nanoparticles

4.1 Relative masses and moles

Key points

- The masses of atoms are compared by measuring them relative to atoms of carbon-12.
- You can work out the relative formula mass of a compound by adding up the relative atomic masses of the elements in it, in the ratio shown by its formula.
- **(H)** One mole of any substance is its relative formula mass, in grams.
- **(H)** Number of moles = $\dfrac{mass\ (g)}{A_r}$ or $= \dfrac{mass\ (g)}{M_r}$
- **(H)** The Avogadro constant is 6.02×10^{23} per mole.

Synoptic links

For more information on isotopes and standard form, see Topic C1.7.

Calculating the relative formula mass

Calculate the relative formula mass of calcium sulfate, $CaSO_4$.

A_r Ca = 40, S = 32, O = 16

M_r $CaSO_4 = 40 + 32 + (16 \times 4) = \mathbf{136}$

Key words: relative atomic mass (A_r), relative formula mass (M_r), mole, Avogadro constant

On the up

You should be able to explain why the relative atomic mass of some elements is not a whole number.

- We use the carbon atom of $^{12}_{6}C$ as a standard atom and compare the masses of all other atoms with $^{12}_{6}C$.

- The **relative atomic mass (A_r)** takes into account the proportions of any isotopes of the element found naturally. So it is the mean (average) relative mass of the atoms of an element compared with the standard carbon atom (which is assigned a relative value of exactly 12). Therefore the relative atomic mass is not always a whole number, for example, the A_r of chlorine is 35.5.

1   Why is the relative atomic mass of chlorine not a whole number?

Calculating relative atomic mass

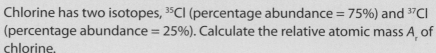

Chlorine has two isotopes, ^{35}Cl (percentage abundance = 75%) and ^{37}Cl (percentage abundance = 25%). Calculate the relative atomic mass A_r of chlorine.

$$A_r \text{ of Cl} = \frac{(35 \times 75) + (37 \times 25)}{100} = \mathbf{35.5}$$

- The **relative formula mass (M_r)** of a substance is found by adding up the A_r of the atoms in its formula. In molecular substances, such as CO_2, you can refer to the relative formula mass as the relative molecular mass.

2   What is the relative formula mass of sodium hydroxide, NaOH?
    A_r: Na = 23, O = 16, H = 1

The mole

- Chemical amounts are measured in moles (use mol as the symbol).

- One **mole** of a substance is the relative atomic mass or relative formula mass of that substance expressed in grams.

- A mole of any substance always contains the same number of atoms, molecules, or ions. This number is the **Avogadro constant**, 6.02×10^{23} per mole.

- To calculate the number of moles you can use these equations:

$$number\ of\ moles = \frac{mass\ (g)}{A_r} \quad or \quad number\ of\ moles = \frac{mass\ (g)}{M_r}$$

You can rearrange the equations to calculate the mass of a certain number of moles.

$$mass\ (g) = number\ of\ moles \times A_r$$

or

$$mass\ (g) = number\ of\ moles \times M_r$$

3   How many moles are present in 2.2 g of carbon dioxide, CO_2? $M_r\ CO_2 = 44$

Higher

4.2 Equations and calculations

Key points

- Balanced symbol equations tell you the number of moles of substances involved in a chemical reaction.
- You can use balanced symbol equations to calculate the masses of reactants and products in a chemical reaction.

Calculating masses from chemical equations

Chemical equations show the reactants and products of a reaction. When they are balanced they show the amounts of atoms, molecules, or ions in the reaction.

For example: $2Mg + O_2 \rightarrow 2MgO$

This shows that two atoms of magnesium react with one molecule of oxygen to form two magnesium ions and two oxide ions.

If we work in moles, the equation tells us that two moles of magnesium atoms react with one mole of oxygen molecules to produce two moles of magnesium oxide.

This means that 48 g of magnesium react with 32 g of oxygen to give 80 g of magnesium oxide. (A_r of Mg = 24, A_r of O = 16)

Alternatively, if we work in relative masses from the equation:
$(2 \times A_r$ of Mg$) + (2 \times A_r$ of O$)$ gives $(2 \times M_r$ of MgO$)$

Converting this to grams it becomes:

(2×24) g Mg + (2×16) g O gives (2×40) g MgO

or 48 g Mg + 32 g O gives 80 g MgO (which is the same as when we used moles).

If we have 5 g of magnesium, we can work out the mass of magnesium oxide it will produce using ratios:

1 g Mg will produce $\dfrac{80}{48}$ g MgO

so 5 g Mg will produce $5 \times \dfrac{80}{48}$ g MgO = 8.33 g of MgO

If we use moles the calculation can be done like this:
1 mole of Mg produces 1 mole of MgO

5 g Mg = $\dfrac{5}{24}$ mole of magnesium and so it will produce $\dfrac{5}{24}$ mole of MgO.

The mass of $\dfrac{5}{24}$ mole of MgO = $\dfrac{5}{24} \times 40$ g

$= \mathbf{8.33\ g}$ of MgO

1 The balanced chemical equation for the reaction between calcium and oxygen is: $2Ca + O_2 \rightarrow 2CaO$

How many moles of calcium atoms react with one mole of oxygen molecules?

2 Calculate the mass of calcium oxide that can be made from 10 g of calcium carbonate in the reaction:
$CaCO_3 \rightarrow CaO + CO_2$ (A_r of Ca = 40, A_r of O = 16, A_r of C = 12)

4.3 From masses to balanced equations

Key points

- You can deduce balanced symbol equations from the masses (and hence the ratio of the numbers of moles) of substances involved in a chemical reaction.

- The reactant that gets used up first in a reaction is called the limiting reactant. This is the reactant that is **not** in excess.

- Therefore, the amounts of product formed in a chemical reaction are determined by the limiting reactant.

Synoptic link

Topic C1.2 gives information about the Law of conservation of mass.

Key word: limiting reactant

- You can calculate the balancing numbers (called multipliers) in an equation from the masses of the substances involved in a reaction. First, calculate the number of moles of each reactant and product. The simplest whole-number ratio of moles of the reactants and products gives you the balanced equation.

1   0.2 mol of hydrogen, H_2, reacts with 0.2 mol of chlorine, Cl_2, to produce 0.4 mol of hydrogen chloride, HCl. What is the equation for the reaction?

Using moles to balance equations

Magnesium, Mg, burns in oxygen gas, O_2, to form magnesium oxide, MgO. When 9.6 g of magnesium is heated until its mass is constant, 16.0 g of magnesium oxide is produced.

a   What mass of oxygen must have been given off in the reaction?

b   Calculate the number of moles of each reactant and product.

c   Show how these can be used to produce the balanced symbol equation for the reaction of magnesium with oxygen. (A_r values: Mg = 24, O = 16)

a   The Law of conservation of mass shows that:

total mass of reactants = total mass of products

So if the mass of oxygen is x g:

magnesium + oxygen → magnesium oxide

     9.6 g   + x g →   16.0 g

x g = 16.0 g – 9.6 g

so mass of oxygen = **6.4 g**

b   You will need to calculate the relative formula masses:

M_r of O_2 = 32 M_r of MgO = 24 + 16 = 40

Then calculate the number of moles using the equation:

$$\text{number of moles} = \frac{\text{mass (g)}}{M_r} \text{ or } \frac{\text{mass (g)}}{A_r}$$

moles of Mg = $\frac{9.6}{24}$ = **0.4**   moles of O_2 = $\frac{6.4}{32}$ = **0.2**   moles of MgO = $\frac{16.0}{40}$ = **0.4**

c   So the ratio = 0.4 mol : 0.2 mol : 0.4 mol

Simplest ratio = 2 mol : 1 mol : 2 mol

So the balanced equation is:

$2Mg + O_2 → 2MgO$

- Usually when you carry out a chemical reaction in experiments, you use an excess of one of the reactants to make sure that the other reactant is all used up.

- The reactant that gets used up first in a reaction is called the **limiting reactant**.

2   What is a limiting reactant?

C4

4.4 The yield of a chemical reaction

Key points

- The yield of a chemical reaction describes how much product is made.
- The percentage yield of a chemical reaction tells you how much product is made compared with the maximum amount that could be made (100%).
- Factors affecting the yield of a chemical reaction include:
  - product being left behind in the apparatus
  - reversible reactions not going to completion
  - some reactants may produce unexpected reactions
  - losses in separating the products from the reaction mixture.

- The **yield** of a chemical process is how much you actually make. The **percentage yield** compares the amount made with the maximum amount that could be made, calculated as a percentage.

Higher

Calculating percentage yield

The percentage yield is calculated using this equation:

$$\text{percentage yield} = \frac{\text{mass of product produced}}{\text{maximum mass of product possible}} \times 100$$

The maximum amount of product possible is calculated from the balanced equation for the reaction.

For example: A student collected 2.3 g of magnesium oxide from 2.0 g of magnesium.

Theoretically: $2Mg + O_2 \rightarrow 2MgO$

so 48 g of Mg should give 80 g of MgO

and so 2.0 g of Mg should give $2 \times \frac{80}{48} = 3.33$ g of MgO.

Percentage yield $= \frac{2.3}{3.33} \times 100$

$\qquad\qquad = \mathbf{69\%}$

> 1   A student made 4.4 g of calcium oxide from 4.0 g of calcium. Calculate the percentage yield. (A_r of Ca = 40, A_r of C = 12, A_r of O = 16)

Synoptic link

For more information about reversible reactions, see Topics C8.6–C8.9.

- When you carry out chemical reactions it is not usually possible to collect the amounts calculated from the chemical equations. Reactions may be reversible and not go to completion, other reactions may happen, and some product may be lost when it is separated or collected from the apparatus.

Key words: yield, percentage yield

> 2   Why is it not usually possible to get 100% yield from a chemical reaction?

On the up

You can achieve the top grades if you are able to calculate the percentage yield of a reaction when given the mass of a reactant used, the mass of product obtained, and the balanced equation for the reaction.

- Using reactions with high yields in industry helps to conserve resources and to reduce waste. Chemical processes should also waste as little energy as possible.

- Working in these ways helps to reduce pollution and makes production more sustainable.

> 3   Why should chemical manufacturers use reactions with high yields?

4.5 Atom economy

- The **atom economy** of a reaction is a measure of the amount of starting materials (the reactants) that end up as the desired product.

- To calculate the percentage atom economy, you use the following equation:

$$\text{percentage atom economy} = \frac{\text{relative formula mass of the desired product from the equation}}{\text{sum of the relative formula masses of the reactants from the equation}} \times 100$$

1   What does atom economy measure?

- It is important to use reactions with a high atom economy. These reactions are more sustainable as they use fewer natural resources. They also give fewer waste products, so there will be fewer costs for waste treatment and disposal.

2   Why are reactions with a high atom economy more sustainable?

Key points

- It is important to maximise atom economy in industrial processes, to conserve the Earth's resources and minimise pollution.

- The atom economy of a reaction uses its balanced equation to compare the relative formula mass of the desired product with the sum of the relative formula masses of the reactants. It is usually expressed as a percentage.

Synoptic links

Topics C14.1 and C14.5 give more information about the Earth's natural resources and the impact of industrial processes on the environment.

To find out more about the extraction of metals, see Topic C5.3.

Study tip

Use bullet points to summarise why high atom economies and high yields are important in industrial processes.

Key word: atom economy

Calculating the percentage atom economy

Lead can be extracted from lead oxide using carbon. The equation for the reaction is:

$$2PbO + C \rightarrow 2Pb + CO_2$$

Calculate the percentage atom economy for this process.

$$\text{percentage atom economy} = \frac{\text{relative formula mass of the desired product from the equation}}{\text{sum of the relative formula masses of the reactants from the equation}} \times 100$$

$$= \frac{M_r (2Pb)}{M_r (2PbO) + M_r (C)} \times 100$$

$$= \frac{(2 \times 207)}{[2 \times (207 + 16)] + 12} \times 100$$

$$= \frac{414}{458} \times 100$$

$$= 90.4\%$$

4.6 Expressing concentrations

Key points

- Concentrations of solutions are measured in grams per decimetre cubed (g/dm³)

- To calculate the mass of solute in a certain volume of solution of known concentration use the equation:

$$\text{concentration (g/dm}^3\text{)} = \frac{\text{mass of solute (g)}}{\text{volume of solution (dm}^3\text{)}}$$

Key word: concentration

Look at the three glasses of orange squash below:

The orange squash is getting less concentrated going left to right (the darker colour indicates more squash is in the same volume of its solution)

- **Concentrations** of solutions can be measured in grams per decimetre cubed (g/dm³).

- If you know the mass of a substance dissolved in a given volume of solution you can calculate its concentration.

  If you are working in dm³ use the equation:

  $$\text{concentration (g/dm}^3\text{)} = \frac{\text{mass of solute (g)}}{\text{volume of solution (dm}^3\text{)}}$$

  If you are working in cm³ use the equation:

  $$\text{concentration (g/dm}^3\text{)} = \frac{\text{mass of solute (g)}}{\text{volume of solution (cm}^3\text{)}} \times 1000$$

Calculating the concentration of a solution

A solution was made by dissolving 0.8 g of sodium hydroxide in 200 cm³ of water. What is the concentration in g/dm³ of the solution?

$$\text{concentration (g/dm}^3\text{)} = \frac{\text{mass of solute (g)}}{\text{volume of solution (cm}^3\text{)}} \times 1000$$

$$= \frac{0.8}{200} \times 1000$$

$$= \textbf{4 g/dm}^3$$

1 4.0 dm³ of solution was made using 2.8 g of potassium hydroxide. What is its concentration in g/dm³?

C4 4.7 Titrations

Key points

- A titration is used to measure accurately what volumes of acid and alkali react together completely.

- The point at which a reaction between an acid and an alkali is complete is called the end point of the reaction.

- You use an acid/base indicator to show the end point of the reaction between an acid and an alkali.

- **(H)** You can calculate the concentration of a solution in mol/dm³, given the mass of solute in a certain volume.

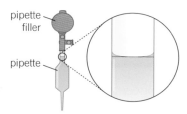

pipette filler

pipette

A pipette and pipette filler

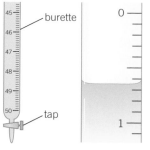

burette

tap

A burette – use the bottom of the meniscus to read the scale. The reading here is 0.65 cm³

Synoptic link

You can find out more about the reactions of acids and bases in Topics C5.6–C5.8

Key words: titration, end point, pipette, burette

- You can use **titrations** to find the exact volumes of acid and alkali that react with each other.

- The point at which the acid and alkali have completely reacted is called the **end point** of the reaction.

Carrying out a titration

You can carry out a titration using the following method.

1. Use a **pipette** to measure a known volume of alkali.

2. Put the alkali into a conical flask.

3. Add a few drops of acid/base indicator to the alkali in the conical flask.

4. Fill a **burette** with acid.

5. Record the initial reading on the burette.

6. Add a small amount of acid to the alkali in the flask. Mix the solutions.

7. Repeat step 6 until the indicator changes colour. This is the end point of the reaction when the acid and alkali have completely reacted.

8. Record the final reading on the burette.

9. Repeat the whole process at least three times. Try to get two results within 0.1 cm³ of each other (called concordant results), and then take the mean of these two results.

---

1 Which piece of apparatus would you use to measure a fixed volume of a solution?

Higher

Concentrations of solutions

Concentrations of solutions are measured in grams per decimetre cubed (g/dm³) or moles per decimetre cubed (mol/dm³).

You can calculate the concentration of a solution in mol/dm³, given the mass of solute in a certain volume.

You can work out the number of moles using the equation:

$$\text{number of moles} = \frac{\text{mass (g)}}{M_r}$$

and then:

- calculate the number of moles in 1 cm³ of solution

- calculate the number of moles in 1000 cm³ (1 dm³) of solution.

---

2 100 cm³ of solution was made using 1.2 g of lithium hydroxide, LiOH. Calculate its concentration in mol/dm³.

4.8 Titration calculations

Key points

- You can use titration to find the unknown concentration of a solution.

- You need to know the accurate concentration of one solution. Then, once the end point is established, the balanced equation gives you the number of moles in a certain volume of solution.

- This value is multiplied up to give the concentration in moles per decimetre cubed (which you can convert to grams per decimetre cubed if necessary).

- Titrations are used to find the volumes of solutions that react exactly.

- If the concentration of one of the solutions is known, and the volumes that react together are known, you can calculate the concentration of the other solution. You can use this information to find the amount of a substance in a sample.

- You can calculate the concentrations using balanced symbol equations and moles.

Titration calculations

A student found that 25.0 cm³ of sodium hydroxide solution with an unknown concentration reacted with exactly 20.0 cm³ of 0.50 mol/dm³ hydrochloric acid. What was the concentration of the sodium hydroxide solution?

The equation for this reaction is: $NaOH(aq) + HCl(aq) \rightarrow NaCl(aq) + H_2O(l)$

The concentration of the HCl is 0.50 mol/dm³, so 0.50 mol of HCl is dissolved in 1000 cm³ of acid.

Therefore 20.0 cm³ of acid contains $20 \times \dfrac{0.50}{1000}$ mol = 0.010 mol HCl

The equation for the reaction tells us that 0.010 mol of HCl will react with exactly 0.010 mol of NaOH.

This means that there must have been 0.010 moles of NaOH in the 25.0 cm³ of solution in the conical flask.

So, the concentration of NaOH solution = $\dfrac{0.010}{25} \times 1000 = \textbf{0.40 mol/dm}^3$

- Notice that, in the answer to the question above, the concentration (0.40 mol/dm³) is given to 2 significant figures. This is to be consistent with the data given to the least number of significant figures in the question (that is, the 2 significant figures in 0.50 mol/dm³).

1   15.0 cm³ of hydrochloric acid reacted exactly with 25.0 cm³ of sodium hydroxide solution that had a concentration of 0.10 mol/dm³.
Calculate the concentration of the hydrochloric acid in mol/dm³.

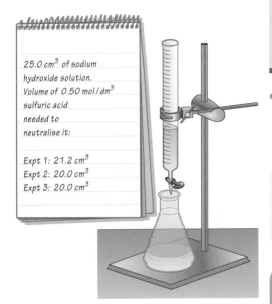

25.0 cm³ of sodium hydroxide solution.
Volume of 0.50 mol/dm³
sulfuric acid
needed to
neutralise it:

Expt 1: 21.2 cm³
Expt 2: 20.0 cm³
Expt 3: 20.0 cm³

From results like these, you can calculate the unknown concentration of a solution – in this case the sodium hydroxide solution. In titrations you usually put the alkaline solution in the conical flask and the dilute acid in the burette

On the up

If you're aiming for the highest grades, you should be able to use given data to do multi-step calculations, giving your answer to an appropriate number of significant figures.

C4

4.9 Volumes of gases

- At the same temperature and pressure, equal numbers of moles of any gas occupy the same volume.

- The volume occupied by 1 mole of any gas is 24 dm³ (24 000 cm³) at room temperature and pressure (20 °C and 1 atmosphere). 24 dm³ per mol is known as the molar gas volume.

- You can use this equation to calculate the number of moles in a given volume of gas:

$$\text{number of moles of gas} = \frac{\text{volume of gas (dm}^3)}{24 \text{ dm}^3}$$

Key points

- A certain volume of gas always contains the same number of gas molecules under the same conditions.

- The volume of 1 mole of any gas at room temperature and pressure is 24 dm³ (24 000 cm³).

- You can use the molar gas volume and balanced symbol equations to calculate volumes of gaseous reactants or products.

1 How many moles of oxygen are present in 48 dm³ of gas?

Calculating the volume of a gas from its mass

What volume of gas is occupied by 2.2 g of carbon dioxide? M_r CO$_2$ = 44

Work out how many moles of CO$_2$ are present:

$$\text{number of moles CO}_2 = \frac{\text{mass (g)}}{M_r}$$

$$= \frac{2.2}{44} = 0.05 \text{ moles}$$

Convert moles into volume.

1 mole of gas occupies 24 dm³

$$\text{number of moles of gas} = \frac{\text{volume of gas (dm}^3)}{24 \text{ dm}^3}$$

Rearranging the equation gives:

volume of gas (dm³) = number of moles of gas × 24 dm³

So, 0.05 moles occupy 0.05 × 24 = **1.2 dm³**

Study tip

You should always include units in your answer when calculating volumes and masses.

Calculating volumes of gaseous reactants or products

You can use a balanced symbol equation to find volumes of reactants and products in reactions involving gases.

Because the same number of moles of gas occupies the same volume, the ratio of the numbers of moles in the balanced equation will be the same as the ratio of the volume of gases involved.

For example,

$$H_2(g) + Cl_2(g) \rightarrow 2HCl(g)$$
1 mole 1 mole   2 moles

The ratio of moles of H$_2$(g) : Cl$_2$(g) : HCl(g) is 1 : 1 : 2

so if you have 100 cm³ of H$_2$(g), the ratio of gases will be

100 cm³ H$_2$: 100 cm³ Cl$_2$: 200 cm³ HCl

2 5 moles of oxygen gas are needed to burn 1 mole of propane gas. Calculate the volume of oxygen needed to burn 2 dm³ of propane gas.

1. Why is an indicator used in an acid/base titration? [1 mark]

2. What is the relative formula mass of sulfuric acid, H_2SO_4? (A_r of H = 1, A_r of S = 32, A_r of O = 16) [2 marks]

3. Give two advantages of using reactions with a high atom economy in industry. [2 marks]

4. Ⓗ 9.20 g of sodium completely reacted with chlorine gas. 23.4 g of sodium chloride was produced. (A_r of Na = 23, A_r of Cl = 35.5)

   a What mass of chlorine reacted with sodium? [1 mark]

   b What is the balanced equation for the reaction? [4 marks]

5. Ⓗ How many moles are present in 8.0 g of magnesium oxide? (A_r of Mg = 24, A_r of O = 16) [2 marks]

6. The equation for the reaction between ethene and hydrogen chloride is:

   $C_2H_4 + HCl \rightarrow C_2H_5Cl$

   Why is the atom economy for the reaction 100%? [1 mark]

7. Ⓗ Calculate the percentage yield if 67.2 g of calcium oxide was made by decomposing 150 g of calcium carbonate. (A_r of Ca = 40, A_r of C = 12, A_r of O = 16) [4 marks]

8. What is the concentration of a solution containing 1.12 g of potassium hydroxide, KOH, in 250 cm^3 of water:

   a in g/dm^3? [1 mark]

   b Ⓗ in mol/dm^3? [3 marks]

9. Ⓗ 200 cm^3 of nitrogen reacted with oxygen. The equation for the reaction is:

   $N_2 + 2O_2 \rightarrow 2NO_2$

   a What volume of oxygen is required to completely react with 200 cm^3 of nitrogen? [1 mark]

   b What volume of nitrogen dioxide is produced? [1 mark]

10. Ⓗ 20 cm^3 of a solution of hydrochloric acid reacted exactly with 25 cm^3 of a 0.10 mol/dm^3 solution of potassium hydroxide. What is the concentration of the hydrochloric acid solution? [4 marks]

Chapter checklist

Tick when you have:

reviewed it after your lesson ✓ ☐ ☐

revised once – some questions right ✓ ✓ ☐

revised twice – all questions right ✓ ✓ ✓

Move on to another topic when you have all three ticks

4.1 Relative masses and moles ☐ ☐ ☐

4.2 Equations and calculations ☐ ☐ ☐

4.3 From masses to balanced equations ☐ ☐ ☐

4.4 The yield of a chemical reaction ☐ ☐ ☐

4.5 Atom economy ☐ ☐ ☐

4.6 Expressing concentrations ☐ ☐ ☐

4.7 Titrations ☐ ☐ ☐

4.8 Titration calculations ☐ ☐ ☐

4.9 Volumes of gases ☐ ☐ ☐

01   An atom of sodium is represented as $^{23}_{11}$Na.

01.1 What is the atomic number of sodium?   [1 mark]

01.2 What is the mass number of sodium?   [1 mark]

01.3 How many protons does a sodium atom contain?   [1 mark]

01.4 How many neutrons does a sodium atom contain?   [1 mark]

01.5 The sodium atom contains 11 electrons. What is the electronic
structure of a sodium atom?   [1 mark]

01.6 Give the name and number of the group in the periodic table
that sodium is in.   [2 marks]

01.7 Sodium metal reacts with chlorine gas to produce solid
sodium chloride. Write the word equation for the reaction.   [1 mark]

01.8 Copy and balance the symbol equation for the reaction.
Include state symbols in the equation.   [2 marks]
$Na(\ ) + Cl_2(\ ) \rightarrow NaCl(\ )$

02   Lithium reacts with fluorine to produce lithium fluoride:
$2Li(s) + F_2(g) \rightarrow 2LiF(s)$

02.1 For each substance in the equation choose the type of bonding
it has from this list: *ionic, covalent, metallic*.   [3 marks]

02.2 Which of the substances in the equation is made from small
molecules?   [1 mark]

02.3 Which of the substances in the equation would you expect
to conduct electricity when solid?   [1 mark]

02.4 Which of the substances in the equation would you expect
to conduct electricity when molten?   [2 marks]

02.5 A lithium atom can be represented as shown in **Figure 1**.

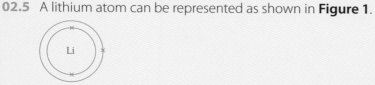

Figure 1

Draw a similar diagram to show a fluorine atom.   [1 mark]

02.6 Use similar diagrams to those in **02.5** to show the bonding in
lithium fluoride.   [4 marks]

02.7 Lithium nanoparticles are used in some batteries.
What is meant by the term nanoparticle?   [1 mark]

02.8 Suggest **one** advantage of using lithium nanoparticles instead
of normal-sized lithium particles.   [1 mark]

03   Mendeleev placed the elements in the periodic table in order of
atomic mass.

03.1 How are the elements ordered in the modern periodic table?   [1 mark]

03.2 Give **two** examples of pairs of elements that would be in the wrong
order if they were arranged in order of their relative atomic masses.   [2 marks]

03.3 How is the atomic number of an element related to the number
of electrons in its atoms?   [1 mark]

03.4 Explain the link between the electrons in an atom and its group
in the periodic table.   [2 marks]

03.5 Explain why using relative atomic masses does not put all the
elements into their correct groups.   [3 marks]

03.6 Why did Mendeleev use atomic mass to order the elements?   [1 mark]

> ## Study tip
>
> When drawing dot and cross
> diagrams to show bonding as in
> question **02**, you only need to
> show the outer electrons of the
> atoms or ions.

04 The element carbon has several forms.

04.1 Diamond is one form of carbon. Describe the structure and bonding in diamond. [4 marks]

04.2 Fullerenes are another form of carbon. Buckminsterfullerene, C_{60}, was the first fullerene to be discovered.
Give **two** similarities between the bonding in buckminsterfullerene and diamond. [2 marks]

04.3 Give **three** differences between the structures of buckminsterfullerene and diamond. [3 marks]

04.4 State **two** uses of fullerenes. [2 marks]

05 Magnesium chloride has the formula $MgCl_2$.

05.1 Calculate the relative formula mass (M_r) of $MgCl_2$. [2 marks]

05.2 What mass of magnesium chloride will be produced when 6.0 g of magnesium reacts with an excess of chlorine? [3 marks]
The equation for the reaction is: $Mg(s) + Cl_2(g) \rightarrow MgCl_2(s)$

05.3 Explain why an ionic substance such as magnesium chloride will conduct electricity when it is molten. [2 marks]

05.4 The structure of magnesium chloride can be represented by the ball and stick model shown in **Figure 2**.

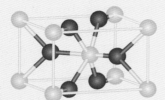

Figure 2

The ball and stick model is not a true representation of the structure of an ionic compound. Give **one** reason why. [1 mark]

06 Students investigated the volume of sulfuric acid that reacted with 25 cm³ of a solution of sodium hydroxide.

06.1 Outline a practical method the students could use to carry out the investigation. [6 marks]

06.2 The students used 25 cm³ of a solution of 0.100 moles per dm³ sodium hydroxide to find the volume of sulfuric acid that reacted. The results are shown in **Table 1**.

Table 1

Titration number	1	2	3	4
Volume of sulfuric acid solution in cm³	15.55	15.05	14.95	15.00

One of the results is anomalous. Which result is anomalous? [1 mark]

06.3 Calculate the mean value for the volume of sulfuric acid needed to neutralise 25 cm³ of 0.100 mol/dm³ sodium hydroxide. [2 marks]

06.4 Ⓗ The equation for the reaction is: $2NaOH + H_2SO_4 \rightarrow Na_2SO_4 + 2H_2O$
Calculate the concentration in mol/dm³ of sulfuric acid in the solution. [4 marks]

06.5 A different student used a different concentration of sodium hydroxide. The solution contained 0.6 g of sodium hydroxide in 25 cm³ of water. What is the concentration of sodium hydroxide in g/dm³? [2 marks]

2 Chemical reactions and energy changes

In the early 19th century, people began experimenting with chemical reactions in a systematic way, organising their results logically. Gradually they began to predict exactly what new substances would be formed and used this knowledge to develop a wide range of different materials and processes. They could extract important resources from the Earth, for example, by using electricity to decompose ionic substances. This is how reactive metals such as aluminium and sodium were discovered.

Energy changes are also an important part of chemical reactions. Transfers of energy take place due to the breaking and formation of bonds. The heating or cooling effects of reactions are used in a range of everyday applications.

I already know...

I will revise...

how to define acids and alkalis in terms of neutralisation reactions.	how to represent neutralisation using an ionic equation.
how to use the pH scale for measuring acidity and alkalinity.	how to calculate the concentration of hydrogen ions in a solution given its whole number pH value.
about displacement reactions and the reactions of acids with metals to produce a salt plus hydrogen.	how to interpret displacement reactions and the reaction between an acid and a metal in terms of reduction and oxidation.
the reactions of acids with alkalis to produce a salt plus water.	how to calculate the concentration of an unknown acid or alkali from experimental results.
combustion and rusting are examples of oxidation reactions.	how to identify and describe oxidation and reduction reactions in terms of electron transfer.
that chemical reactions are exothermic and endothermic.	how to use bond energy values to calculate the approximate energy change accompanying a reaction.

5.1 The reactivity series

Key points

- The metals can be placed in order of reactivity by their reactions with water and dilute acid.
- Hydrogen gas is given off if metals react with water or dilute acids. The gas 'pops' with a lighted spill.

Synoptic link

Topic C2.3 gives information about the reactions of the alkali metals in Group 1.

Key words: oxidised, reduced, ore, reactivity series

Study tip

When a lighted splint is applied to hydrogen, the gas burns with a squeaky 'pop'.

- Most metals are found as compounds in rocks. Many of these metals have been **oxidised** (i.e., have oxygen added) by oxygen in the air to form their oxides. So to extract the metals, the metal oxide must be **reduced** (i.e., have oxygen removed).
- Rock that contains enough of a metal or a metal compound to make it economically worthwhile extracting the metal is called an **ore**.
- The **reactivity series** is a list of metals in order of their reactivity, with the most reactive metals at the top and the least reactive metals at the bottom.
- The most reactive metals form positive ions most easily.

1 What is an ore?

Metals plus water

The Group 1 metals react vigorously with water, giving off hydrogen gas and leaving alkaline hydroxide solutions. For example,

potassium + water → potassium hydroxide + hydrogen

$$2K(s) + 2H_2O(l) \rightarrow 2KOH(aq) + H_2(g)$$

- Magnesium metal only reacts very slowly with cold water. Less reactive metals such as copper do not react at all with water.

2 Using the reactions of copper, magnesium, and potassium with water, place these metals in order of reactivity.

Metals plus dilute acid

- Hydrogen gas is produced if metals react with water or dilute acid.
- Metals such as magnesium react with acids to produce a salt and hydrogen gas.
- The rate at which the metal gives off hydrogen gas can be used to judge the relative reactivity of magnesium, zinc, and iron.
- The table below summarises the reactions of some important metals with water and dilute acid. It shows the reactivity series of metals.

Order of reactivity	Reaction with water	Reaction with dilute acid
potassium	fizz, giving off hydrogen, leaving an alkaline solution of metal hydroxide	explode
sodium		
lithium		
calcium		fizz, giving off hydrogen and forming a salt
magnesium	very slow reaction with cold water	
zinc		
iron		
copper	no reaction	no reaction
silver		
gold		

5.2 Displacement reactions

Key points

- A more reactive metal will displace a less reactive metal from its aqueous solution.
- The non-metals hydrogen and carbon can be given positions in the reactivity series on the basis of displacement reactions.
- **(H)** Oxidation is the loss of electrons.
- **(H)** Reduction is the gain of electrons.

Synoptic link

See Topic C14.4 for more about an industrial application of the displacement of copper ions from solution by iron.

Key words: displacement reaction, ionic equation, oxidation, reduction, half equation

Study tip

You can use the phrase **OILRIG** to remember the definition of oxidation and reduction reactions:
Oxidation **I**s **L**oss of electrons
Reduction **I**s **G**ain of electrons

On the up

You should be able to predict which metals react with metal salt solutions and explain why the reaction takes place.

- A more reactive metal will displace a less reactive metal from the aqueous solution of one of its salts.
- For example, look at the **displacement reaction** below.

magnesium + copper(II) sulfate →
magnesium sulfate + copper
$Mg(s) + CuSO_4(aq) \rightarrow MgSO_4(aq) + Cu(s)$

Here is the **ionic equation** for this displacement reaction (showing only the ions that change in the reaction). **H**
$Mg(s) + Cu^{2+}(aq) \rightarrow Mg^{2+}(aq) + Cu(s)$

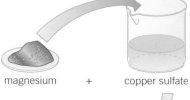
magnesium + copper sulfate

magnesium sulfate + copper

Magnesium displaces copper from copper(II) sulfate solution

1 Write a word and balanced symbol equation, including state symbols, for the reaction between zinc granules and copper sulfate solution.

- You can include the non-metals hydrogen and carbon in the reactivity series by using displacement reactions. Hydrogen comes between copper and lead in the series. Carbon is between aluminium and zinc.

2 Place the following five elements in order of reactivity, with the most reactive first.
hydrogen, magnesium, copper, iron, carbon

Oxidation is the loss of electrons.

Reduction is the gain of electrons.

You can use the displacement reaction between zinc and copper sulfate to explain a redox (reduction–oxidation) reaction.

Ionic equation: $Zn(s) + Cu^{2+}(aq) \rightarrow Zn^{2+}(aq) + Cu(s)$

You can use **half equations** to show what happens to each reactant in the ionic equation:

$Zn(s) \rightarrow Zn^{2+}(aq) + 2e^-$

Zinc atoms lose two electrons to form zinc ions. This is oxidation (the loss of electrons). We say that zinc atoms have been oxidised.

These two electrons from zinc are gained by the copper ions as they form copper atoms.

$Cu^{2+}(aq) + 2e^- \rightarrow Cu(s)$

This is reduction (the gain of electrons). The copper ions have been reduced.

3 In the displacement reaction between magnesium atoms and copper ions in solution, which reactant is oxidised and which is reduced?

Higher

5.3 Extracting metals

Key points

- A metal ore contains enough of the metal to make it economic to extract the metal. Ores are mined from the ground and might need to be concentrated before the metal is extracted and purified.
- Gold and other unreactive metals can be found in their native state.
- The reactivity series helps you to decide the best way to extract a metal from its ore. The oxides of metals below carbon in the series can be reduced by carbon to give the metal element.
- Metals more reactive than carbon cannot be extracted from their ores using carbon. They are extracted by electrolysing the molten metal compound.

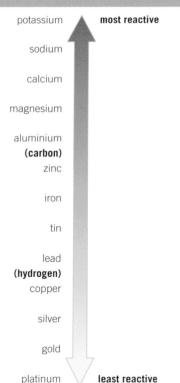

This reactivity series shows the position of the non-metals carbon and hydrogen in the series

- Mining ores often involves digging up large amounts of rock. The metal ore may need to be concentrated before the metal is extracted. These processes can produce large amounts of waste and may have major impacts on the environment.

- Whether it is worth extracting a particular metal depends on:
  - how easy it is to extract it from its ore
  - how much metal the ore contains
  - changing demand for the metal.

- A few unreactive metals, such as gold, that are low in the reactivity series, are found in the Earth as the metal. Gold can be separated from rocks by physical methods.

- However, most metals are found as compounds. These metals have to be extracted by chemical reactions.

1 Why is gold found in the Earth as a metal not a compound?

- Metals can be extracted from compounds by displacement using a more reactive element.

- Metals that are less reactive than carbon can be extracted from their oxides by heating with carbon. A reduction reaction takes place as carbon removes the oxygen from the oxide to produce the metal. This method is used commercially if possible.

- For example,

$$\text{lead oxide} + \text{carbon} \xrightarrow{\text{heat}} \text{lead} + \text{carbon dioxide}$$
$$2PbO(s) + C(s) \rightarrow 2Pb(l) + CO_2(g)$$

2 a Name two metals that have oxides that can be reduced by carbon.
  b What do we call the removal of oxygen from a metal oxide?

- The metals that are more reactive than carbon are not extracted from their ores by reduction with carbon. Instead they are extracted by **electrolysis** of the molten metal compound.

Key word: electrolysis

Synoptic link

To find out more about the use of electrolysis to extract reactive metals, see Topic C6.3.

Study tip

Make a list of metals in order of reactivity.

Student Book
pages 90–91
C5

5.4 Salts from metals

- Acids will react with metals that are above hydrogen in the reactivity series.

- However, the reactions of acids with very reactive metals, such as sodium and potassium, are too violent to carry out safely.

- When metals react with acids, they produce a **salt** and hydrogen gas.

  acid   + metal →   a salt   + hydrogen

  $H_2SO_4(aq) + Zn(s) \rightarrow ZnSO_4(aq) + H_2(g)$

1   Name a metal other than zinc that can safely react with an acid to produce a salt.

- The salt produced depends on the metal and acid you use. Magnesium metal always makes salts containing magnesium ions, Mg^{2+}.

  Salts produced from:

  - hydrochloric acid, HCl, are **chlorides** (containing Cl^- ions)

  - sulfuric acid, H_2SO_4, are **sulfates** (containing SO_4^{2-} ions)

  - nitric acid, HNO_3, are **nitrates** (containing NO_3^- ions).

2   Name the salt produced from magnesium and hydrochloric acid.

Explaining the reaction between a metal and an acid

The equation for the reaction between magnesium and dilute sulfuric acid is:

$Mg(s) + H_2SO_4(aq) \rightarrow MgSO_4(aq) + H_2(g)$

The ionic equation is:

$Mg(s) + 2H^+(aq) \rightarrow Mg^{2+}(aq) + H_2(g)$

You can use half equations to show what happens to each reactant in the ionic equation.

$$Mg(s) \rightarrow Mg^{2+}(aq) + 2e^-$$

Magnesium atoms lose two electrons to form magnesium ions. These two electrons from magnesium are gained by two hydrogen ions from the acid. A molecule of hydrogen gas, H_2, is formed.

$$2H^+(aq) + 2e^- \rightarrow H_2(g)$$

Electrons have been transferred in the reaction:

  - The magnesium atoms have lost electrons, so magnesium atoms have been **oxidised**.

  - The hydrogen ions have gained electrons, so hydrogen ions have been **reduced**.

The reaction of a metal with an acid is always a redox reaction.

Higher

Key points

- A salt is a compound formed when the hydrogen in an acid is wholly or partially replaced by metal or ammonium ions.

- Salts can be made by reacting a suitable metal with an acid. The metal must be above hydrogen in the reactivity series, but not dangerously reactive.

- The reaction between a metal and an acid produces hydrogen gas as well as a salt. A sample of the salt made can then be crystallised out of solution by evaporating off the water.

- **(H)** The reaction between a metal and an acid is an example of a redox reaction. The metal atoms lose electrons and are oxidised, and hydrogen ions from the acid gain electrons and are reduced.

Key word: salt

Study tip

Learn the formulae of the negative ions, chloride Cl^-, sulfate SO_4^{2-}, and nitrate NO_3^- to help you write the formulae of salts.

5.5 Salts from insoluble bases

Key points

- When an acid reacts with a base, a neutralisation reaction occurs.

- The reaction between an acid and a base produces a salt and water.

- The sum of the charges on the ions in a salt adds up to zero. This enables you to work out the formula of salts, knowing the charges on the ions present.

- A pure, dry sample of the salt made in an acid–base reaction can be crystallised out of solution by evaporating off most of the water, and drying with filter papers if necessary.

Key word: neutralisation

On the up

You should be able to describe a method to prepare a pure, dry sample of a soluble salt from an acid and an insoluble base or carbonate.

- Salts contain positive metal ions (or ammonium ions, NH_4^+) and a negative ion from an acid.

- Salts have no overall charge, because the sum of the charges on their ions equals zero. So once you know the charges on the ions that make up a salt, you can work out its formula.

Formulae of salts

What is the formula of the salt magnesium nitrate?

The formulae of the ions are Mg^{2+} and NO_3^-.

The magnesium ion has a 2+ charge so you will need a 2– charge, from two NO_3^- ions, to balance the charges.

Therefore Mg^{2+} and NO_3^- must combine in the ratio 1 : 2, making the formula of the salt, magnesium nitrate, $Mg(NO_3)_2$.

- Metal oxides and metal hydroxides are bases. When an acid reacts with a base, a **neutralisation** reaction takes place, and a salt and water are produced.

$$\text{acid} + \text{base} \rightarrow \text{a salt} + \text{water}$$
$$2HCl(aq) + MgO(s) \rightarrow MgCl_2(aq) + H_2O(l)$$

1 Why do you add excess of the base when making a salt?

Making a copper salt

You can make copper sulfate crystals from copper(II) oxide (an insoluble base) and sulfuric acid. The equation for the reaction is:

$$\text{sulfuric acid} + \text{copper(II) oxide} \rightarrow \text{copper(II) sulfate} + \text{water}$$
$$H_2SO_4(aq) + CuO(s) \rightarrow CuSO_4(aq) + H_2O(l)$$

1

2

warm gently

Add insoluble copper oxide to sulfuric acid and stir. Warm gently on a tripod and gauze (do not boil).

The solution turns blue as the reaction occurs, showing that copper sulfate is being formed. Excess black copper oxide can be seen.

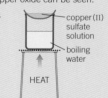

3

4
copper(II) sulfate solution
boiling water
HEAT

When the reaction is complete, filter the solution to remove excess copper oxide.

You can evaporate the water so that crystals of copper sulfate start to form. Stop heating when you see the first crystals appear at the edge of the solution. Then leave for the rest of the water to evaporate off slowly. This will give you larger crystals. Any small excess of solution on the crystals can be removed by dabbing between filter papers (do not touch the solution), then leaving to dry.

5.6 Making more salts

Key points

- An indicator is needed when a soluble salt is prepared by reacting an alkali with an acid.

- The titration can be repeated without the indicator to make a salt, and then a pure, dry sample of its crystals can be prepared.

- A carbonate reacts with an acid to produce a salt, water, and carbon dioxide gas.

Synoptic link

Titrations are covered in Topic C4.7.

- You can make soluble salts by reacting an acid and an alkali, for example,

$$acid + alkali \rightarrow salt + water$$
$$HCl(aq) + NaOH(aq) \rightarrow NaCl(aq) + H_2O(l)$$

- You can represent the neutralisation reaction between an acid and an alkali by this ionic equation:

$$H^+(aq) + OH^-(aq) \rightarrow H_2O(l)$$

- There is no visible change when acids react with alkalis so you need to use an indicator or a pH meter to show when the reaction is complete. The solid salt can be obtained from the solution by crystallisation.

1   What compound is produced in every neutralisation reaction?

- You can also make soluble salts by reacting an acid and a carbonate, for example,

acid + a carbonate → a salt + water + carbon dioxide

For calcium carbonate, the reaction with hydrochloric acid is:

calcium carbonate + hydrochloric acid → calcium chloride + water + carbon dioxide
$$CaCO_3(s) + 2HCl(aq) \rightarrow CaCl_2(aq) + H_2O(l) + CO_2(g)$$

Powdered limestone is used to raise the pH of acidic soils or lakes affected by acid rain, making use of the reaction between calcium carbonate and acid

2   Write a word equation and a balanced symbol equation, including state symbols, for the reaction of magnesium carbonate with hydrochloric acid.

Student Book
pages 96–97

C5

5.7 Neutralisation and the pH scale

Key points

- Acids are substances that produce H⁺(aq) ions when you add them to water.
- Bases are substances that will neutralise acids.
- An alkali is a soluble hydroxide. Alkalis produce OH⁻(aq) ions when you add them to water.
- You can use the pH scale to show how acidic or alkaline a solution is.
- Solutions with pH values less than 7 are acidic, pH values more than 7 are alkaline, and a pH value of 7 indicates a neutral solution.

Key words: neutral, acid, base, alkali, pH scale

Study tip

Practise using the symbols > and < to describe the pH of acids and alkalis.

- Pure water is **neutral** and has a pH value of 7.
- **Acids** are substances that produce hydrogen ions, H⁺(aq), when they are added to water.
- When you dissolve a substance in water you make an aqueous solution.
- The state symbol (aq) shows that the ions are in aqueous solution. Excess hydrogen ions make solutions acidic and they have pH values of less than 7.

1 When you add acids to water, which ions are produced?

- **Bases** react with acids and neutralise them. Metal oxides and metal hydroxides are bases.
- **Alkalis** are bases that dissolve in water to make alkaline solutions. Alkalis are soluble hydroxides and produce hydroxide ions, OH⁻(aq), in the solution. Alkaline solutions have pH values greater than 7.

2 What is an alkali?

- The **pH scale** has values from 0 to 14. Solutions that are very acidic have low pH values. Solutions that are very alkaline have high pH values.
- Indicators have different colours in acidic and alkaline solutions. Universal indicator (UI) has a range of colours at different pH values.

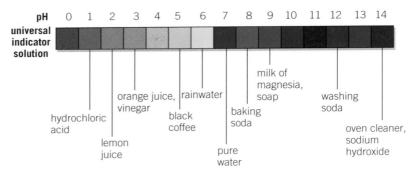

The pH scale tells you how acidic or alkaline a solution is

3 Which indicator can tell you the pH of a solution?

Maths skills

You can use the mathematical symbols '>' (read as 'is greater than') and '<' ('is less than') when interpreting pH values.

You can say:

pH < 7 indicates an acidic solution, i.e., pH values less than 7 are acidic.

pH > 7 indicates an alkaline solution, i.e., pH values greater than 7 are alkaline.

5.8 Strong and weak acids

Key points

- Aqueous solutions of weak acids, such as carboxylic acids, have a higher pH value than solutions of strong acids with the same concentration.

- As the pH decreases by one unit, the hydrogen ion concentration of the solution increases by a factor of 10 (i.e., one order of magnitude).

Synoptic links

For more information on ethanoic acid see Topic C10.2.

For more information about reversible reactions at equilibrium, see Topic C8.8.

Key words: strong acid, equilibrium, weak acid

- In aqueous solution, hydrochloric acid ionises completely to hydrogen ions and chloride ions.

$$HCl(aq) \rightarrow H^+(aq) + Cl^-(aq)$$

- Acids that ionise completely in aqueous solutions are known as **strong acids**.

- When ethanoic acid dissolves in water, it does not ionise completely and some of the ethanoic acid molecules remain as molecules in the solution.

$$CH_3COOH(aq) \rightleftharpoons CH_3COO^-(aq) + H^+(aq)$$

A position of **equilibrium** is reached in which both whole molecules and their ions are present in the solution.

- Acids that do not ionise completely in aqueous solution are known as **weak acids**.

- In aqueous solutions of equal concentration, weak acids have a higher pH and react more slowly than strong acids.

---

1 Write a balanced equation to show that ethanoic acid is a weak acid.

How are pH values related to the concentration of H⁺(aq) ions?

- As hydrochloric acid is diluted, the concentration of $H^+(aq)$ ions decreases and the pH of the solution increases. The table below shows the pattern.

Concentration of H⁺(aq) ions in mol/dm³	pH value
0.1	1.0
0.01	2.0
0.001	3.0
0.0001	4.0

- So as the concentration of $H^+(aq)$ ions increases by a factor of 10, the pH value decreases by one unit.

---

2 $H^+(aq)$ ions have a concentration of 0.0001 mol/dm³.
   Write this value in standard form.

3 A solution has a pH value of 5.0.
   What is the concentration of $H^+(aq)$ ions?

1 What is the name for a rock from which metal can be extracted economically? [1 mark]

2 Why is gold found in the Earth as the metal? [1 mark]

3 What process is used to extract very reactive metals, e.g., aluminium from aluminium oxide? [1 mark]

4 Dilute hydrochloric acid is added to sodium hydroxide solution.
   a What type of substance is sodium hydroxide? [1 mark]
   b What type of reaction happens? [1 mark]
   c How can you show when the reaction is complete? [2 marks]
   d Write a word equation for the reaction. [1 mark]

5 Which ion do acids produce when they dissolve in water? [1 mark]

6 Zinc reacts with hydrochloric acid. Name the products formed in the reaction. [2 marks]

7 What is meant by oxidation? [1 mark]

8 a What type of chemical reaction is used to extract lead from lead oxide? [1 mark]
   b Write a word equation for the reaction. [1 mark]

9 Outline a method you can use to make copper sulfate crystals from copper oxide and dilute sulfuric acid. [6 marks]

10 Ⓗ What is the difference between strong and weak acids? [2 marks]

11 Magnesium metal reacts with copper sulfate solution. Write an equation, including state symbols, for the reaction. [2 marks]

12 Ⓗ The ionic equation for the reaction between magnesium and dilute sulfuric acid is:

$$Mg(s) + 2H^+(aq) \rightarrow Mg^{2+}(aq) + H_2(g)$$

   a Explain how magnesium atoms form magnesium ions. [2 marks]
   b What type of reaction is taking place? [1 mark]

13 Ⓗ A 0.001 mol/dm³ solution of a strong acid has a pH of 3.0.
   What is the concentration of the acid when the pH is 2? [1 mark]

Chapter checklist

Tick when you have:

reviewed it after your lesson	✓	☐	☐
revised once – some questions right	✓	✓	☐
revised twice – all questions right	✓	✓	✓

Move on to another topic when you have all three ticks

5.1 The reactivity series ☐ ☐ ☐
5.2 Displacement reactions ☐ ☐ ☐
5.3 Extracting metals ☐ ☐ ☐
5.4 Salts from metals ☐ ☐ ☐
5.5 Salts from insoluble bases ☐ ☐ ☐
5.6 Making more salts ☐ ☐ ☐
5.7 Neutralisation and the pH scale ☐ ☐ ☐
5.8 Strong and weak acids ☐ ☐ ☐

Student Book
pages 102–103

C6

6.1 Introduction to electrolysis

Key points

- Electrolysis breaks down a substance using electricity.
- Ionic compounds can only be electrolysed when they are molten or in water. This is because their ions are then free to move around and carry their charge to the electrodes.
- In electrolysis, positive ions move to the cathode (negative electrode), while negative ions move to the anode (positive electrode).

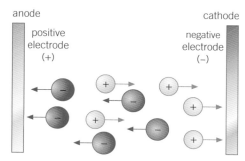

An ion always moves towards the oppositely charged electrode

Synoptic links

You can find information about ionic bonding and the properties of ionic compounds in Topics C3.3 and C3.4.

Key words: electrolysis, electrolyte, inert, anode, cathode

- **Electrolysis** is the process that uses electricity to break down ionic compounds into elements.
- When electricity passes through a molten ionic compound or a solution containing ions, electrolysis takes place.
- The substance that is broken down is called the **electrolyte**.

1   What must be done to ionic compounds before they can be electrolysed?

- The electrical circuit has two electrodes that make contact with the electrolyte.
- The electrodes are often made of an **inert** substance that does not react with the products.
- The positive electrode is called the **anode**. The negative electrode is called the **cathode**.
- During electrolysis, positively charged ions move to the cathode (negative electrode). At the same time, the negative ions move to the anode (positive electrode), as opposite charges attract. When the ions reach the electrodes, they lose their charge and become atoms or molecules.
- When electrolysed, molten ionic compounds produce the metal element at the cathode and the non-metal element at the anode.

2   Molten zinc chloride is electrolysed. Name the substance produced:

   a   at the anode (positive electrode)

   b   at the cathode (negative electrode).

- When electrolysing ionic compounds in solution, water also forms ions, so it is more difficult to predict what will be formed. Positively charged ions form hydrogen or the metal, depending on the reactivity of the metal. Negatively charged ions form oxygen (and water) or halogens.

Study tip

Remember – never PANIC if you can't remember the charge on each electrode!
Positive **A**node, **N**egative **I**s **C**athode

On the up

You should be able to explain why electrolysis can only occur when an ionic compound is molten or in solution.

6.2 Changes at the electrodes

Key points

- In electrolysis, the ions move towards the oppositely charged electrodes.
- At the negative electrode (cathode), positive ions gain electrons, so are reduced.
- At the positive electrode (anode), negative ions lose their extra electrons, so are oxidised.
- When electrolysis happens in aqueous solution, the less reactive element, either hydrogen or the metal, is usually produced at the cathode. At the anode, you get either:
  - oxygen gas given off (plus water), from discharged hydroxide ions present in aqueous solutions, or
  - a halogen produced if the electrolyte is a solution of a halide.

Synoptic link

See Topic C5.2 for more information on reduction and oxidation in terms of electron transfer.

Key word: half equations

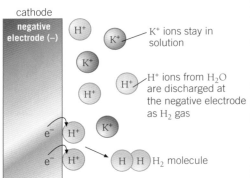

cathode
negative electrode (–)

K⁺ ions stay in solution

H⁺ ions from H₂O are discharged at the negative electrode as H₂ gas

H H H₂ molecule

Here is the cathode in the electrolysis of a solution of a potassium compound

---

- When positively charged ions reach the cathode (negative electrode), they gain electrons to become neutral atoms. Ions with a single positive charge gain one electron and those with a 2+ charge gain two electrons.
- At the anode (positive electrode), negative ions lose electrons to become neutral atoms. Some non-metal atoms combine to form molecules, for example, bromine forms Br_2.

1 What happens at the cathode (negative electrode) when sodium ions become sodium atoms?

Half equations

You can represent the changes at the electrodes by **half equations**. The half equations for lead bromide are:

at the cathode (negative electrode): $Pb^{2+} + 2e^- \rightarrow Pb$
The positive ions gain electrons, so they are reduced.

at the anode (positive electrode): $2Br^- \rightarrow Br_2 + 2e^-$
The negative ions lose electrons, so they are oxidised.

2 Copy and complete the half equation for the formation of chlorine at an anode:

$$2Cl^- \rightarrow \_\_ + \_\_$$

The effect of water

- In the electrolysis of aqueous solutions, because water contains hydrogen ions, $H^+(aq)$, and hydroxide ions, $OH^-(aq)$, different products can be obtained.
- At the negative electrode (cathode), if two elements can be produced, the less reactive element will usually be formed. When aqueous solutions are electrolysed, the positively charged metal ions and $H^+(aq)$ ions are attracted to the negative electrode. Hydrogen gas is produced at the negative electrode if hydrogen is less reactive than the metal.

At the cathode (–):
$2H^+(aq) + 2e^- \rightarrow H_2(g)$

- At the positive electrode (anode), oxygen gas is usually produced from aqueous solutions.

At the anode (+):
$4OH^-(aq) \rightarrow O_2(g) + 2H_2O(l) + 4e^-$

- However, if the solution contains halide ions, then a halogen will be produced.

3 Name the product(s) at **a** the anode (positive electrode), and **b** the cathode (negative electrode) when aqueous potassium sulfate solution is electrolysed.

6.3 The extraction of aluminium

Key points

- Aluminium oxide, from the ore bauxite, is electrolysed in the extraction of aluminium metal.
- The aluminium oxide is mixed with molten cryolite to lower its melting point, reducing the energy needed to extract the aluminium.
- Aluminium forms at the cathode (negative electrode) and oxygen forms at the anode (positive electrode).
- The carbon anodes are replaced regularly as they gradually burn away as the oxygen reacts with the hot carbon anodes, forming carbon dioxide gas.

Synoptic link

You can see aluminium's position in the reactivity series in Topic C5.3.

- You get aluminium from the ore bauxite, which is mainly aluminium oxide.
- Aluminium is more reactive than carbon and so it must be extracted by electrolysis.
- The electrolysis of aluminium oxide requires large amounts of energy. Aluminium oxide melts at over 2000 °C. Aluminium oxide is mixed with cryolite, which lowers the melting point of the mixture. The molten mixture can then be electrolysed at 850 °C.
- Aluminium metal is produced at the negative electrode and oxygen gas at the positive electrode.

1   Why is aluminium oxide mixed with cryolite in the electrolysis cell?

- The overall reaction in the electrolysis cell is:

aluminium oxide → aluminium + oxygen

$$2Al_2O_3(l) \longrightarrow 4Al(l) + 3O_2(g)$$

- Fresh aluminium oxide is added as aluminium and oxygen are produced.
- At the cathode (negative electrode) aluminium ions gain electrons and form aluminium atoms. The molten aluminium metal is collected from the bottom of the cell.
- At the anode (positive electrode) oxide ions lose electrons to form oxygen atoms. The oxygen atoms then form oxygen molecules.

2   What are the products formed at each electrode in the electrolysis cell?

Half equations for the process

At the cathode (negative electrode) reduction occurs:

$$Al^{3+}(l) + 3e^- \rightarrow Al(l)$$

At the anode (positive electrode) oxidation occurs:

$$2O^{2-}(l) \rightarrow O_2(g) + 4e^-$$

- The oxygen that forms at the hot carbon anodes (negative electrodes) reacts to produce carbon dioxide gas:

$$C(s) + O_2(g) \rightarrow CO_2(g)$$

This means that the carbon anodes gradually burn away and have to be replaced regularly.

Higher

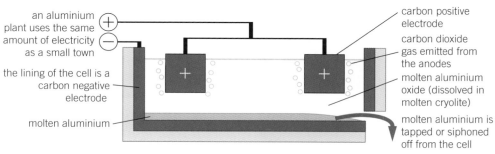

A cell used in the extraction of aluminium by electrolysis

6.4 Electrolysis of aqueous solutions

- Brine is a solution of sodium chloride in water. The equation for the electrolysis of brine is:

$$\text{sodium chloride solution} \xrightarrow{\text{electrolysis}} \text{hydrogen gas + chlorine gas + sodium hydroxide solution}$$

- Sodium chloride solution contains sodium ions, $Na^+(aq)$, chloride ions, $Cl^-(aq)$, hydrogen ions, $H^+(aq)$, and hydroxide ions, $OH^-(aq)$.

- Chloride ions, $Cl^-(aq)$, and hydroxide ions, $OH^-(aq)$, are attracted to the positive electrode. Because the solution contains halide ions, the chloride ions are discharged. Each chloride ion loses one electron to form a chlorine atom. Two chlorine atoms bond to form a chlorine molecule and chlorine gas is produced.

The half equation is: **H**

$$2Cl^-(aq) \rightarrow Cl_2(g) + 2e^-$$

- Hydrogen ions, $H^+(aq)$, and sodium ions, $Na^+(aq)$, are attracted to the negative electrode. Hydrogen ions are discharged, because sodium metal is more reactive than hydrogen. Each hydrogen ion gains one electron to form a hydrogen atom. Two hydrogen atoms bond to form a hydrogen molecule and hydrogen gas is produced.

The half equation is: **H**

$$2H^+(aq) + 2e^- \rightarrow H_2(g)$$

- When we electrolyse brine, hydrogen is produced at the cathode (negative electrode) from the hydrogen ions. Chlorine is produced at the anode (positive electrode) from the chloride ions.

1   Why is hydrogen produced when sodium chloride solution is electrolysed?

- This leaves an alkaline solution of sodium ions and hydroxide ions, $NaOH(aq)$.

Investigating the electrolysis of a solution

You can investigate the electrolysis of aqueous solutions, such as sodium chloride solution, using an electrolysis cell.

Pass an electric current through the cell until the tubes are approximately half full of gas. This prevents the gases being inhaled.

You can test the gases produced to identify which gases are present.

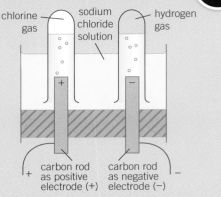

2   Name the three useful products formed when aqueous potassium chloride solution is electrolysed.

1. What is the positive electrode called? [1 mark]

2. What type of ions move to the negative electrode? [1 mark]

3. Name the products at the positive and negative electrodes when the following molten salts are electrolysed:
   a calcium chloride [2 marks]
   b lithium iodide [2 marks]
   c magnesium bromide [2 marks]

4. What type of compounds can be broken down by electrolysis? [1 mark]

5. Which ions from water are present in the solution when aqueous solutions are electrolysed? [2 marks]

6. Aluminium is manufactured from aluminium oxide, Al_2O_3.
   a Why is the electrolysis of aluminium oxide an expensive process? [2 marks]
   b Name the electrolyte in the process. [1 mark]
   c Why is cryolite used in the process to make aluminium? [1 mark]
   d Why does the positive electrode have to be regularly replaced in this process? [4 marks]

7. a Predict the three products formed when aqueous sodium chloride solution is electrolysed. [3 marks]
   b Describe an investigation to show how you could test your prediction. [6 marks]

8. Explain as fully as you can what happens at the electrodes when molten sodium chloride is electrolysed. [6 marks]

9. Ⓗ Write half equations for the reactions at the a positive, and b negative electrodes when molten lead bromide is electrolysed. [2 marks]

10. An aqueous solution of potassium sulfate is electrolysed.
    a Explain why hydrogen is produced at the negative electrode. [3 marks]
    b Ⓗ Write half equations for the reactions occurring at the positive and negative electrodes. [2 marks]

Chapter checklist

✓

Tick when you have:

Tick when you have:				6.1 Introduction to electrolysis			
reviewed it after your lesson	✓	☐	☐	6.2 Changes at the electrodes	☐	☐	☐
revised once – some questions right	✓	✓	☐	6.3 Extraction of aluminium	☐	☐	☐
revised twice – all questions right	✓	✓	✓	6.4 Electrolysis of aqueous solutions	☐	☐	☐

Move on to another topic when you have all three ticks

7.1 Exothermic and endothermic reactions

Key points

- Energy is conserved in chemical reactions. It is neither created nor destroyed.
- A reaction in which energy is transferred from the reacting substances to their surroundings is called an exothermic reaction.
- A reaction in which energy is transferred to the reacting substances from their surroundings is called an endothermic reaction.

Synoptic links

Neutralisation reactions are covered in Topic C5.7, and combustion reactions are covered in Topic C9.3.

Key words: exothermic, endothermic

- When chemical reactions take place, energy is transferred as bonds are broken and made. Reactions that transfer energy to the surroundings are called **exothermic** reactions. The energy transferred often heats up the surroundings and so the temperature increases.
- Exothermic reactions include:
  - combustion, such as burning fuels
  - oxidation reactions, such as respiration
  - neutralisation reactions involving acids and bases.
- In an exothermic reaction, the products have less energy content than the reactants.

1 How can you tell from observations that burning natural gas is an exothermic reaction?

- **Endothermic** reactions take in energy from the surroundings. In endothermic reactions, the products have more energy content than the reactants.
- Some endothermic reactions cause a decrease in temperature and others require a supply of energy to keep the reaction going.
- Thermal decomposition reactions are examples of endothermic reactions.

2 What are the two ways that show that a reaction is endothermic?

Investigating temperature changes

You can use a poly(styrene) cup and a thermometer to investigate the energy changes in reactions involving at least one solution.

Record the initial temperatures of any solutions and the maximum or minimum temperature reached in the course of the reaction.

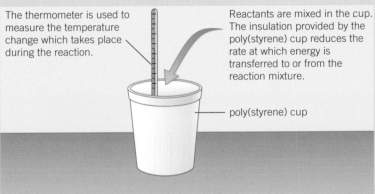

The thermometer is used to measure the temperature change which takes place during the reaction.

Reactants are mixed in the cup. The insulation provided by the poly(styrene) cup reduces the rate at which energy is transferred to or from the reaction mixture.

poly(styrene) cup

3 State two ways in which you could make the data you collect more accurate.

7.2 Using energy transfers from reactions

- Exothermic reactions can be used to heat things.

- Hand warmers and self-heating cans use exothermic reactions. In some hand warmers and cans the reactants are used up and so they cannot be used again. They use reactions such as the oxidation of iron or the reaction of calcium oxide with water. Other hand warmers use a reversible reaction such as the crystallisation of a salt. Once used, the pack can be heated in boiling water to re-dissolve the salt. These can be reused many times.

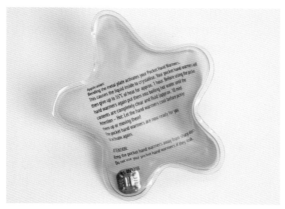

A reusable hand warmer based on recrystallisation

1 Suggest one advantage and one disadvantage of a reusable hand warmer compared with a single-use hand warmer.

- Endothermic changes can be used to cool things.

- Some chemical cold packs contain ammonium nitrate and water that are kept separated. When mixed together the ammonium nitrate dissolves and takes in energy from the surroundings. The cold pack can be used on sports injuries or to cool drinks. The reaction is reversible but not in the pack and so this type of pack can be used only once.

Instant cold packs can be applied to injuries

2 Suggest one advantage and one disadvantage of a chemical cold pack.

7.3 Reaction profiles

- **Reaction profile** diagrams show the relative amounts of energy contained in the reactants and the products. A curved line, drawn from reactants to products, shows the progress of the reaction.

- For an exothermic reaction the products are at a lower energy level than the reactants. This means that when the reactants form the products, energy is transferred to the surroundings. The surroundings get hotter.

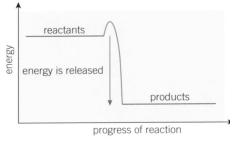

The reaction profile for an exothermic reaction

1   What is the difference in energy levels between products and reactants in an exothermic reaction?

- In an endothermic reaction the products are at a higher energy level than the reactants. As the reactants react to form products, energy is transferred from the surroundings to the reaction mixture. The temperature of the surroundings decreases, because energy is taken in during the reaction. The surroundings get colder.

- The difference in energy between the reactants and the peak of the curve indicates the energy input required for the reaction to take place. This is called the **activation energy**. The activation energy is the minimum energy needed for the reaction to happen.

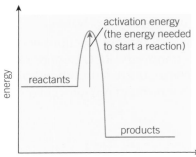

This reaction profile shows the activation energy for an exothermic reaction

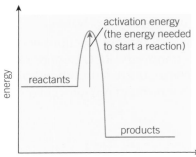

The reaction profile for an endothermic reaction

2   Draw an energy level diagram for an endothermic reaction.

Bond breaking and bond making

During a chemical reaction, bonds in the reactants must be broken for the reaction to happen. Breaking bonds is endothermic because energy is taken in from the surroundings.

When new bonds in the products are formed, energy is released and transferred to the surroundings. Making bonds is an exothermic process.

$$H\text{---}H + O\text{---}O \longrightarrow \underset{H}{\overset{H}{>}}O + \underset{H}{\overset{H}{>}}O$$

Hydrogen and oxygen react to make water. The bonds between hydrogen atoms and between oxygen atoms in their molecules have to be broken so that bonds between oxygen atoms and hydrogen atoms in water molecules can be formed

Higher

7.4 Bond energy calculations

Key points

- In chemical reactions, energy must be supplied to break the bonds between atoms in the reactants.
- When new bonds are formed between atoms in a chemical reaction, energy is released.
- In an exothermic reaction, the energy released when new bonds are formed is greater than the energy required when bonds are broken.
- In an endothermic reaction, the energy released when new bonds are formed is less than the energy needed when bonds are broken.
- You can calculate the overall energy change in a chemical reaction using bond energies.

Key word: bond energy

- Exothermic reactions transfer energy to the surroundings. The energy released when new bonds are formed in the products is more than the energy needed to break the bonds in the reactants.

- Endothermic reactions transfer energy from the surroundings to the reacting chemicals. The energy needed to break the bonds in the reactants is more than the energy released when new bonds are formed in the products.

- The energy needed to break the bond between two atoms is called the **bond energy** for that bond. An equal amount of energy is released when the bond forms between two atoms and so we can use bond energies to calculate the overall energy change for a reaction. Bond energies are measured in kJ/mol.

- You need the balanced equation for the reaction to calculate the energy change for a reaction. Then calculate:
  - the total amount of energy needed to break all of the bonds in the reactants
  - the total amount of energy released in making all of the bonds in the products
  - the difference between the two totals.

Bond	Bond energy in kJ/mol
C—C	347
C—H	413
H—O	464
O=O	498
C=O	805

Calculating the overall energy change using bond energies

Use the bond energies in the table to calculate the energy change for burning methane:

$CH_4 + 2O_2 \rightarrow CO_2 + 2H_2O$

Bonds broken: $4 \times$ C–H $+ 2 \times$ O=O

Energy needed $= (4 \times 413) + (2 \times 498) = 2648$ kJ/mol

Bonds formed: $2 \times$ C=O $+ 4 \times$ H—O

Energy released $= (2 \times 805) + (4 \times 464) = 3466$ kJ/mol

So overall energy change (difference) = 3466 kJ/mol – 2648 kJ/mol = 818 kJ/mol

Energy change for the reaction = **818 kJ/mol** (This energy is released to the surroundings in an exothermic reaction, as more energy is released as bonds form (3466 kJ/mol) than is needed when they are broken (2648 kJ/mol)).

1 Calculate the energy change for burning propane using the bond energies in the table above.

$$C_3H_8 + 5O_2 \rightarrow 3CO_2 + 4H_2O$$

7.5 Chemical cells and batteries

Key points

- Metals tend to lose electrons and form positive ions.

- When two metals are dipped in a salt solution and joined by a wire, the more reactive metal will donate electrons to the less reactive metal. This forms a simple electrical cell.

- The greater the difference in reactivity between the two metals, the higher the voltage produced by the cell.

Synoptic link

The reactivity series of metals is covered in Topics C5.1 and C5.3.

- You can use the difference in reactivity of metals to make electrical cells and batteries. A battery is made up of two or more cells joined together.

- You can make a cell by joining two metals together by a wire and dipping them into a salt solution. In the cell, electrons will flow through the wire from the more reactive metal to the less reactive metal. The flow of electrons is an electric current. The current will flow in the circuit until one of the reactants is used up.

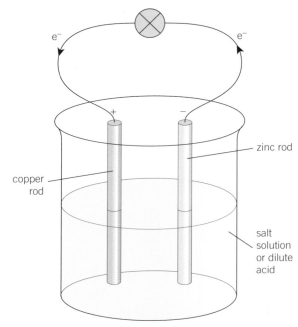

An electrical cell made from zinc and copper. The electrons flow from the more reactive metal (zinc) to the less reactive metal (copper). So zinc acts as the negative terminal of the cell, providing electrons to the external circuit

1 What is the difference between a cell and a battery?

- The greater the difference in reactivity between the two metals used, the higher the voltage produced. The more reactive metal has a greater tendency to give away electrons and form a positive ion.

2 Which way will electrons flow in an electrical cell made from zinc and copper?

- The voltage produced in a cell also depends on the electrolyte.

- In non-rechargeable cells, such as zinc/carbon or alkaline batteries, once one of the reactants has run out, the cell stops working.

- Rechargeable cells can be recharged and used again. In the recharging process, the cell is connected to a power supply that reverses the reactions that occur at each electrode when the cell is discharging. This regenerates the original reactants.

7.6 Fuel cells

- Using hydrogen as a fuel could help reduce global warming because the reaction does not produce carbon dioxide. Hydrogen burns well and the only product is water.

1 What is one advantage of using hydrogen as a fuel?

- A **fuel cell** is an efficient use of the energy from oxidising hydrogen. Fuel cells are fed with hydrogen and oxygen and produce water. Most of the energy released in the reaction is transferred by electricity (electrons moving through the wire). This electricity can be used to do useful work such as running a vehicle.

- In the fuel cell, hydrogen gas is supplied as a fuel to the negative electrode. Hydrogen diffuses through the electrode and reacts with hydroxide ions to form water and provides a source of electrons to an external circuit.

The half equation at the negative electrode is: **H**

$$2H_2(g) + 4OH^-(aq) \rightarrow 4H_2O(l) + 4e^-$$

- Oxygen gas is supplied to the positive electrode. Oxygen diffuses through the electrode and reacts with water to form hydroxide ions, accepting electrons from the external circuit.

The half equation at the positive electrode is: **H**

$$O_2(g) + 2H_2O(l) + 4e^- \rightarrow 4OH^-(aq)$$

- The overall change in the hydrogen fuel cell, is the oxidation of hydrogen (the fuel):

$$2H_2(g) + O_2(g) \rightarrow 2H_2O(l)$$

- The only waste product is water. So hydrogen fuel cells offer a potential alternative to conventional fossil fuels and to rechargeable cells and batteries.

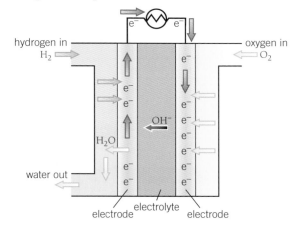

A hydrogen fuel cell, which has an alkaline electrolyte, such as potassium hydroxide solution. Notice that the only waste product is water

- Advantages of hydrogen fuel cells:
  - do not need to be electrically recharged
  - no pollutants are produced
  - can be a range of sizes for different uses.

- Disadvantages of hydrogen fuel cells:
  - hydrogen is highly flammable
  - hydrogen is sometimes produced for the cell by non-renewable sources
  - hydrogen is difficult to store.

2 What is the waste product of the hydrogen fuel cell?

1 What do we call the type of reaction that results in the temperature of its surroundings decreasing? [1 mark]

2 State two advantages of using hydrogen fuel cells. [2 marks]

3 Name one practical use for an endothermic reaction. [1 mark]

4 Name two types of reaction that are exothermic. [2 marks]

5 What is meant by the 'activation energy' of a reaction? [2 marks]

6 **a** Describe how you can make a simple cell. [3 marks]

    b How could you change the cell to increase the voltage? [1 mark]

7 Write a balanced equation for the overall reaction that takes place in a hydrogen fuel cell. [2 marks]

8 How can you tell from a reaction profile diagram that a reaction is endothermic? [1 mark]

9 Why must calcium carbonate be heated continuously to convert it into calcium oxide and carbon dioxide? [2 marks]

10 What determines the reactivity of a metal in terms of ion formation? [2 marks]

11 Draw a reaction profile diagram for the exothermic reaction, $CH_4 + 2O_2 \rightarrow CO_2 + 2H_2O$. Show and label the reaction pathway, the activation energy, and the energy change of the reaction. [4 marks]

12 **H** Calculate the energy change for the reaction, $H_2C{=}CH_2 + 3O_2 \rightarrow 2CO_2 + 2H_2O$ using bond energies: [6 marks]

Bond	Bond energy in kJ/mol
C—H	413
C=C	612
O=O	498
C=O	805
H—O	464

Chapter checklist

Tick when you have:

reviewed it after your lesson ✔ ☐ ☐

revised once – some questions right ✔ ✔ ☐

revised twice – all questions right ✔ ✔ ✔

Move on to another topic when you have all three ticks

7.1 Exothermic and endothermic reactions ☐ ☐ ☐

7.2 Using energy transfers from reactions ☐ ☐ ☐

7.3 Reaction profiles ☐ ☐ ☐

7.4 Bond energy calculations ☐ ☐ ☐

7.5 Chemical cells and batteries ☐ ☐ ☐

7.6 Fuel cells ☐ ☐ ☐

01 Some types of hand warmers use the reaction of iron with oxygen:

$$4Fe(s) + 3O_2(g) \rightarrow 2Fe_2O_3(g)$$

The hand warmers contain iron powder moistened with salt solution. The salt acts as a catalyst. When air is allowed into the mixture it reacts with the iron and the pack gets warm. The hand warmers can work for several hours.

01.1 What type of reaction transfers energy to the surroundings? [1 mark]

01.2 What is meant by a catalyst? [2 marks]

01.3 Suggest **two** ways, other than increasing the temperature, that could be used to make the reaction in the hand warmer go faster. [2 marks]

02 **Table 1** gives information about three substances that can be used as fuels.

Table 1

Name of fuel	Energy produced in kJ/g	Volume of 1 gram at 20 °C in cm³
ethanol	27	0.79
hydrogen	120	12 000.00
petrol	44	0.72

02.1 Which fuel produces the most energy per gram? [1 mark]

02.2 Calculate the energy produced by 1 cm³ of petrol. Give your answer to 2 significant figures. [3 marks]

02.3 Explain the difference in volume between 1 gram of hydrogen and 1 gram of ethanol. [2 marks]

02.4 Suggest **two** advantages of using hydrogen as a fuel. [2 marks]

02.5 Suggest **two** disadvantages of using hydrogen as a fuel in cars. [2 marks]

03 Copper oxide is insoluble. Solid copper oxide reacts with dilute hydrochloric acid to form copper chloride and water.

$$CuO(\ ) + HCl(\ ) \rightarrow CuCl_2(aq) + \underline{\hspace{1cm}}(l)$$

03.1 Copy, complete, and balance the equation for the reaction. Add the missing state symbols. [3 marks]

03.2 What type of substance is copper oxide? [1 mark]

03.3 Describe how you could make a solution of copper chloride from excess copper oxide and hydrochloric acid. [4 marks]

03.4 Why is excess copper oxide added? [1 mark]

03.5 Describe how you could obtain crystals of copper chloride from the copper chloride solution you prepared in question **03.3**. [4 marks]

04.1 Dilute sulfuric acid, H_2SO_4, reacts with sodium hydroxide, NaOH, to produce sodium sulfate solution, Na_2SO_4, and water. Write a balanced chemical equation for the reaction. Include state symbols in your equation. [3 marks]

04.2 Outline a plan to describe how you would find the maximum temperature change when different volumes of sodium hydroxide solution are added to 25 cm³ of sulfuric acid. [6 marks]

04.3 A student carried out the investigation. The student's results were:

Volume of sodium hydroxide added in cm³	Maximum temperature in °C
5	24.0
10	26.5
15	28.5
20	30.0
25	31.0

Study tip

In longer, high-mark questions like question **04.2**, plan your answer by writing brief notes of the main steps, make sure these are in a sensible order, and then write your answer.

30	30.5
35	30.0
40	29.0

What type of reaction is shown by the results? [1 mark]

04.4 Describe two patterns shown by the results.
Explain why the temperature changes in the patterns you described. [4 marks]

05 A student electrolysed sodium chloride solution using the apparatus shown in **Figure 1**.

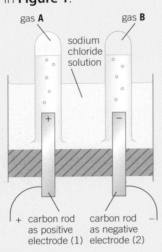

gas **A**   gas **B**

sodium chloride solution

+   −

+ carbon rod   carbon rod −
as positive   as negative
electrode (1)   electrode (2)

Figure 1

05.1 Name gas **A**. [1 mark]
05.2 Name gas **B**. [1 mark]
05.3 Name the product that remains in the solution. [1 mark]
05.4 Explain why sodium metal is not produced in this electrolysis. [2 marks]
05.5 Ⓗ What type of reaction is this? Explain why. [2 marks]
05.6 Ⓗ Write a half equation for the reaction at the positive electrode.
Include state symbols in your equation. [3 marks]

06 **Table 2** shows some bond energies.

Table 2

Bond	Bond energy in kJ/mol
C—H	413
C—C	347
O=O	498
C=O	805
H—O	464

Pentane burns in air. The equation for the reaction is:

$$C_5H_{12} + 8O_2 \rightarrow 5CO_2 + 6H_2O$$

06.1 Use the bond energies in the table to calculate the energy change for this reaction. Show all your working. [5 marks]
06.2 Explain, in terms of bond energies, why this reaction is exothermic. [2 marks]

3 Rates, equilibrium, and organic chemistry

Chemical reactions can occur at vastly different rates and there are many variables that can be manipulated in order to change their speed. Chemical reactions may also be reversible so conditions will affect the yield of desired product. In industry, chemists and chemical engineers determine the effect of different variables on rate of reaction and yield of product.

The great variety of organic compounds is possible because carbon atoms can form chains and rings linked by C—C bonds. Chemists can modify these organic molecules in many ways to make new and useful materials such as polymers, pharmaceuticals, perfumes, flavourings, dyes, and detergents.

I already know...

I already know...	I will revise...
the properties of the different states of matter (solid, liquid and gas) in terms of the particle model, including gas pressure.	how to apply the particle model in the collision theory used to explain the effect of changing conditions on the rate of a reaction.
what catalysts do.	how to explain how catalysts can affect the rate of a reaction in terms of their effect on the activation energy of the reaction, including reaction profile diagrams.
simple techniques for separating mixtures such as distillation.	how fractional distillation is used to separate different fractions from the mixture of hydrocarbons in crude oil.
some examples of combustion and thermal decomposition reactions.	the products of complete and incomplete combustion of fuels from crude oil, and the use of thermal decomposition in the process of cracking large hydrocarbons into smaller, more useful products.
the structure and bonding of some simple molecular substances.	how to draw the displayed formula of alkanes, alkenes, alcohols, carboxylic acids, and esters.
polymers are long molecules made of many repeating groups of atoms.	the different types of bonding between monomers and how this affects the properties of a polymer.

8.1 Rate of reaction

Key points

- You can find out the rate of a chemical reaction by monitoring the amount of reactants used up over time.

- Alternatively, you can find out the rate of reaction by measuring the amount of products made over time.

- The gradient of the line at any given time on the graph drawn from such an experiment tells you the rate of reaction at that time. The steeper the gradient, the faster the reaction.

- Ⓗ To calculate the rate of reaction at a specific time, draw the tangent to the curve, then calculate its gradient.

Key word: gradient

Study tips

The faster the rate, the shorter the time it takes for the reaction to finish. So rate is inversely proportional to time.

The steeper the line on the graph, the faster the rate of reaction.

On the up

You should be able to calculate the mean rate of reaction. To achieve the top grades, you should be able to use a graph to calculate the gradient to measure rate of reaction at any specific time.

- The rate of a reaction measures the speed of a reaction. Reactions happen at different rates. An explosion is a very fast reaction, whereas the rusting of iron is a very slow reaction.

marble chips and hydrochloric acid — cotton wool bung — conical flask — top-pan balance

- To work out the rate of a chemical reaction, you can find out how quickly:
  - the reactants are used up as they make products, or
  - the products of the reaction are made.

- You can find out how quickly the reactants are used up in some reactions by measuring the mass of a reaction mixture.

  If the reaction gives off a gas, the mass of the reaction mixture decreases. You can measure and record the mass at regular time intervals.

- The rate of a reaction at any given time can be found from the **gradient**, or slope, of the line on a graph of amount of reactant or product against time. The steeper the gradient, the faster the reaction is at that time.

- A graph can be produced by measuring the mass of gas released or the volume of gas produced at intervals of time.

- You can calculate the mean rate of reaction after a given time using the equation:

$$\text{mean rate of reaction} = \frac{\text{quantity of reactant used}}{\text{time taken}} \textbf{ or } \frac{\text{quantity of product formed}}{\text{time taken}}$$

1 What two types of measurement must be made to find the average rate of a reaction?

Calculating the rate of reaction at a specific time

You can use a reaction graph to find the rate of the reaction at a given time.

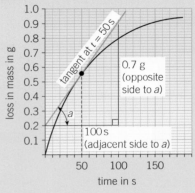

$$\text{Rate at 50 s} = \frac{0.7}{100} = \textbf{0.007 g/s}$$

(The gradient is the tangent of angle *a* in the right-angled triangle, i.e., opposite side divided by adjacent side.)

2 How would the line on the graph differ if you plot 'mass of reacting mixture' on the vertical axis?

8.2 Collision theory and surface area

Key points

- Particles must collide, with a certain minimum amount of energy, before they can react.
- The minimum amount of energy that particles must have in order to react is called the activation energy of a reaction.
- The rate of a chemical reaction increases if the surface area to volume ratio of any solid reactants is increased. This increases the frequency of collisions between reacting particles.

Synoptic link

For more information about the activation energy of a reaction, see Topic C7.3.

Synoptic link

Surface area to volume ratio (in the context of nanoparticles) is covered in Topic C3.11.

Key words: collision theory, activation energy

- The **collision theory** states that reactions can only happen if particles collide. The particles must also collide with enough energy to change into new substances. The minimum energy they need to react is called the **activation energy**.

- You can increase the rate of reaction if you:
  - increase the frequency of reacting particles colliding with each other
  - increase the energy they have when they collide.

1   What do you call the minimum energy needed for particles to react?

- Factors that increase the frequency of collisions, or the energy of the particles, will increase the rate of the reaction. Increasing the temperature, the concentration of reactants in solutions, the pressure of reacting gases, or the surface area of solids will increase the rate of a reaction. Using a catalyst will also increase the rate of a reaction (see Topic C8.5).

- Breaking large pieces of a solid into smaller pieces exposes new surfaces and so increases the surface area (and its surface area to volume ratio). This means there are more collisions in the same time. So a powder reacts faster than an equal mass of large lumps of a substance. The finer the powder, the faster the reaction.

2   Why do powders react faster than large pieces of solid?

Calculating the mean rate of reaction

A group of students timed how long it took before no more gas was given off from calcium carbonate added to excess dilute hydrochloric acid. They performed three experiments using the same volume and concentration of acid, and 2.50 g of large, medium, and then small marble chips. Here are their results:

Size of marble chips	Time until no more bubbles of gas appeared in s
small	102
medium	188
large	294

Calculate the mean rate of reaction of each size of marble chips used.

$$\text{mean rate} = \frac{\text{mass of reactant used up (g)}}{\text{time (s)}}$$

Small chips

$\dfrac{2.50}{102}$

$= 0.0245$ g/s

(fastest mean rate)

Medium chips

$\dfrac{2.50}{188}$

$= 0.0133$ g/s

Large chips

$\dfrac{2.50}{294}$

$= 0.00850$ g/s

(slowest mean rate)

Notice that the data are given to 3 significant figures, so the answers are consistent with the data provided.

8.3 The effect of temperature

Key points

- Reactions happen more quickly as the temperature increases.
- Increasing the temperature increases the rate of reaction because particles collide more frequently and more energetically.
- More of the collisions occurring in a given time result in a reaction, because a higher proportion of particles have energy greater than the activation energy.

Study tip

When explaining rates of reaction, make sure you refer to the <u>rate</u> at which the particles collide. You can describe the collisions as 'more frequent collisions' or that the 'collisions occur more often' or there are 'more collisions in a given time'. However, do not just say there are 'more collisions' as this would be insufficient for a mark.

- Increasing the temperature increases the rate of a reaction. There are two reasons:
  - Particles collide more often.
  - Particles collide with more energy.
- Therefore, a small change in temperature has a large effect on reaction rates. At ordinary temperatures a rise of 10 °C will roughly double the rate of many reactions, so they go twice as fast.

1 Why does a small change in temperature have a large effect on the rate of reaction?

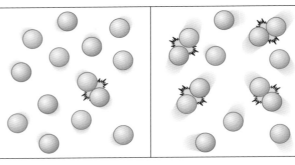

cold – slow movement, less frequent collisions, little energy

hot – fast movement, more frequent collisions, more energy

More frequent collisions, with more energy – both of these factors increase the rate of a chemical reaction caused by increasing the temperature

8.4 The effect of concentration and pressure

Key points

- Increasing the concentration of reactants in solutions increases the frequency of collisions between particles, and so increases the rate of reaction.
- Increasing the pressure of reacting gases also increases the frequency of collisions, and so increases the rate of reaction.

- Increasing the concentration of reactants in a solution increases the rate of reaction because there are more particles of the reactants moving around in the same volume of solution. The more 'crowded together' the reactant particles are, the more likely it is that they will collide. So the increased frequency of collisions results in a faster reaction.

1 Why do reactions in solutions go faster at higher concentrations?

- Increasing the pressure of reacting gases has the same effect. Increased pressure squashes the gas particles more closely together. There are more particles of gas in a given space. This increases the chance that they will collide and react. So increasing the pressure produces more frequent collisions, which will increase the rate of the reaction.

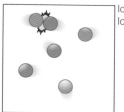

low concentration/
low pressure

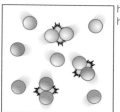

high concentration/
high pressure

Increasing the concentration of solutions or the pressure of gases both mean that particles are closer together. This increases the frequency of collisions between reactant particles, so the reaction rate increases

2   Why does increasing the pressure increase the rate of a reaction of two gases?

Concentration and rate of reaction

You can investigate how changes in concentration affect the rate of reaction using one of the methods below.

Measuring the increasing volume of gas given off

If a reaction produces a gas, you can collect the gas and measure the volume given off at time intervals. You can collect the gas in a gas syringe.

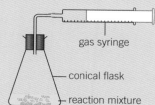

gas syringe

conical flask

reaction mixture

You can also collect the gas in an inverted measuring cylinder, or burette, filled with water.

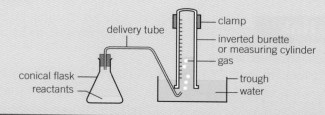

delivery tube

clamp

inverted burette
or measuring cylinder

gas

conical flask

reactants

trough

water

- You can find the rate of a reaction by plotting the volume of gas given off as the reaction progresses over time. Alternatively, you can time how long it takes to collect a fixed volume of gas using the same apparatus.

Measuring the decreasing light passing through a solution

Some reactions in solution make a suspension of an insoluble solid (precipitate). This makes the solution go cloudy. You measure how long it takes until you look through the solution and can no longer see the cross. You can use this to measure the rate at which the precipitate is forming in the reaction.

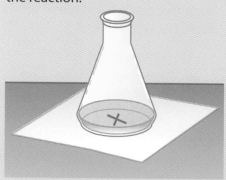

8.5 The effect of catalysts

Key points

- A catalyst speeds up the rate of a chemical reaction, but it is not used up itself during the reaction. It remains chemically unchanged.
- Different catalysts are needed for different reactions.
- Catalysts are used whenever possible in industry to increase rates of reaction and reduce energy costs.

Synoptic link

Reaction profiles and exothermic reactions are covered in Topic C7.3.

Study tip

Catalysts change only the rate of reactions. They do not change the products or yield of a reaction.

- A **catalyst** changes the rate of a reaction. A catalyst is not used up or changed chemically in the reaction, so it can be used over and over again. Different catalysts are needed for different reactions.

- Catalysts increase rates of reaction by providing an alternative reaction pathway to the products, with a lower activation energy than the reaction without the catalyst present. So with a catalyst, a higher proportion of the reactant particles have sufficient energy to react. This means that the frequency of effective collisions (collisions that result in a reaction) increases and the rate of reaction speeds up.

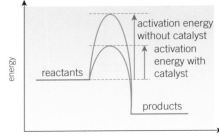

The reaction profile of an uncatalysed and a catalysed exothermic reaction. The catalyst lowers the activation energy of the reaction

1 Why can catalysts be used over and over again?

- Catalysts are used in many industrial processes because they can reduce the energy and the time needed for reactions. This helps to reduce costs and reduce impacts on the environment.

2 What are the benefits of using catalysts in industrial processes?

Key word: catalyst

8.6 Reversible reactions

Key points

- In a reversible reaction, the products of the reaction can react to make the original reactants.
- You can show a reversible reaction using the ⇌ sign.

Key word: reversible reaction

- In some reactions the products can react together to make the original reactants again. This is called a **reversible reaction**.

- A reversible reaction can go in both directions, so two 'half-arrows' are used in the equation. One 'half-arrow' points in the forwards direction and one points in the backwards direction:

$$A + B \rightleftharpoons C + D$$

1 What is a reversible reaction?

- When ammonium chloride is heated, it decomposes to produce ammonia gas and hydrogen chloride gas. When the gases cool down, they react to form ammonium chloride again. This is an example of a reversible reaction:

$$\text{ammonium chloride} \underset{\text{cool}}{\overset{\text{heat}}{\rightleftharpoons}} \text{ammonia} + \text{hydrogen chloride}$$

$$NH_4Cl(s) \rightleftharpoons NH_3(g) + HCl(g)$$

Rates and equilibrium

8.7 Energy and reversible reactions

Blue hydrated copper(II) sulfate and white anhydrous copper(II) sulfate

- In reversible reactions, the forward and reverse reactions involve equal but opposite energy transfers. A reversible reaction that is exothermic in one direction must be endothermic in the other direction. The amount of energy released by the exothermic reaction exactly equals the amount taken in by the endothermic reaction.

- The reaction below shows a reversible reaction where **A** and **B** react to form **C** and **D**. The products of this reaction (**C** and **D**) can then react to form **A** and **B** again.

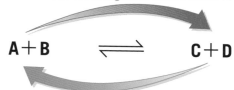

if the reaction **transfers** energy to the surroundings when it goes in this direction ...

$$A+B \rightleftharpoons C+D$$

... it will **take in** exactly the same amount of energy from the surroundings when it goes in this direction.

A reversible reaction

- When you heat blue **hydrated** copper(II) sulfate crystals the reaction is endothermic:

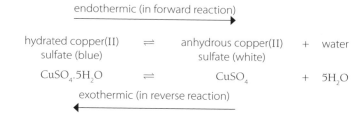

endothermic (in forward reaction) →

| hydrated copper(II) sulfate (blue) | $\rightleftharpoons$ | anhydrous copper(II) sulfate (white) | + | water |
| $CuSO_4.5H_2O$ | $\rightleftharpoons$ | $CuSO_4$ | + | $5H_2O$ |

← exothermic (in reverse reaction)

- When you add water to **anhydrous** copper(II) sulfate, hydrated copper(II) sulfate is formed. The reaction in this direction is exothermic.

1 Why must blue copper(II) sulfate be heated continuously to change it into anhydrous copper(II) sulfate?

2 Why does adding water to anhydrous copper(II) sulfate cause the mixture to get hot?

8.8 Dynamic equilibrium

Key points

- In a reversible reaction, the products of the reaction can react to re-form the original reactants.

- In a closed system, the rate of the forward and reverse reactions is equal at equilibrium.

- Changing the reaction conditions can change the amounts of products and reactants in a reaction mixture at equilibrium.

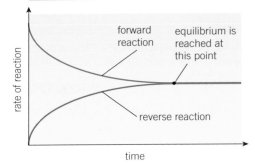

In a reversible reaction at equilibrium, the rate of the forward reaction is the same as the rate of the reverse reaction

Synoptic link

For more information about economic and safety issues associated with reversible reactions in industry, see Topics C15.6 and C15.8.

- In a **closed system** no reactants or products can escape. For a reversible reaction in a closed system, **equilibrium** is reached when the rate of the forward reaction is equal to the rate of the reverse reaction. At equilibrium both reactions continue to happen, but the amounts of reactants and products remain constant.

1   Explain what is meant by equilibrium.

You can change the relative amounts of the reactants and products in a reacting mixture at equilibrium by changing the conditions. This is important for the chemical industry in controlling reactions.

This is an application of **Le Châtelier's Principle**. Le Châtelier noticed that whenever you change the conditions in a system at equilibrium, the position of equilibrium shifts so as to cancel out the change. The change in conditions can be changes in concentration, pressure, or temperature.

For example, increasing the concentration of a reactant will cause more products to be formed as the system tries to achieve equilibrium. If a product is removed, more reactants will react to try to achieve equilibrium and so more product is formed.

For example, for the reversible reaction:

$$ICl + Cl_2 \rightleftharpoons ICl_3$$

If chlorine is added, the concentration of chlorine is increased and more ICl_3 is produced, thereby reducing the concentration of chlorine again.

If chlorine is removed, the concentration of chlorine is decreased and more ICl is produced, thereby increasing the concentration of chlorine again.

2   To make SO_3, the reaction $2SO_2(g) + O_2(g) \rightleftharpoons 2SO_3(g)$ is done in a reactor over a heated catalyst. Why is the SO_3 removed from the reactor as soon as it is made?

Key words: closed system, equilibrium, Le Châtelier's Principle

Higher

Higher tier

8.9 Altering conditions

Key points

- Pressure can affect reversible reactions involving gases at equilibrium. Increasing the pressure favours the reaction that forms fewer molecules of

Changing pressure

- If you change the conditions of a system at equilibrium, the position of equilibrium shifts as if to try to cancel out the change.

- For reversible reactions that have different numbers of molecules of gases on one side of the balanced equation than the other, changing the pressure will affect the position of equilibrium. For example, if the pressure is increased, the position of equilibrium will shift to try to reduce the pressure (favouring the reaction that produces fewer molecules of gas).

gas. Decreasing the pressure favours the reaction that forms the greater number of molecules of gas.

- You can change the relative amount of products formed at equilibrium, by changing the temperature at which you carry out a reversible reaction.

- Increasing the temperature favours the endothermic reaction. Decreasing the temperature favours the exothermic reaction.

Synoptic links

For information on how changes in pressure and temperature are useful in industry, look at Topics C15.5 and C15.6.

- This is summarised in the table:

If the forward reaction produces more molecules of gas …	If the forward reaction produces fewer molecules of gas …
… an increase in pressure decreases the amount of products formed.	… an increase in pressure increases the amount of products formed.
… a decrease in pressure increases the amount of products formed.	… a decrease in pressure decreases the amount of products formed.

- For example, in the reversible reaction: $2NO_2(g) \rightleftharpoons N_2O_4(g)$ there are more gaseous reactant molecules than gaseous product molecules. Therefore increasing the pressure will increase the amount of N_2O_4 (product) in the mixture at equilibrium.

1 For the reaction $2SO_2(g) + O_2(g) \rightleftharpoons 2SO_3(g)$, what change in pressure will increase the amount of SO_3 in the equilibrium mixture?

Changing temperature

- Reversible reactions are exothermic in one direction and endothermic in the other direction.

- Increasing the temperature favours the reaction in the endothermic reaction. The equilibrium shifts as if to lower the temperature by taking in energy.

- Decreasing the temperature favours the exothermic reaction. This is summarised in the table:

If the forward reaction is exothermic …	If the forward reaction is endothermic …
… an increase in temperature decreases the amount of products formed.	… an increase in temperature increases the amount of products formed.
… a decrease in temperature increases the amount of products formed.	… a decrease in temperature decreases the amount of products formed.

- For example, for the reversible reaction: $2NO_2(g) \rightleftharpoons N_2O_4(g)$ the forward reaction is exothermic, so increasing the temperature will produce more NO_2 (reactant) in the mixture at equilibrium.

The effect of changing the temperature on $2NO_2(g) \rightleftharpoons N_2O_4(g)$

2 The reaction $2SO_2(g) + O_2(g) \rightleftharpoons 2SO_3(g)$ is exothermic in the forward direction. What change in temperature will increase the amount of SO_3 at equilibrium?

On the up

To achieve the highest grades, you should be able to describe and explain the effect of changing the temperature and pressure on a reversible reaction at equilibrium.

1 What does the sign $\rightleftharpoons$ signify in a reaction? [1 mark]

2 Iron reacts with dilute sulfuric acid. Given the same mass of iron, which form of iron will react fastest when acid is added and which will react slowest?

iron wool   iron filings   iron nails   a block of iron [2 marks]

3 What is a catalyst? [2 marks]

4 State two pieces of apparatus that could be used to measure the volume of a gas given off as a reaction takes place. [2 marks]

5 Sodium thiosulfate and hydrochloric acid produce a cloudy precipitate when they react.

a What is a precipitate? [2 marks]

b Describe a practical method to find the rate of reaction. [6 marks]

6 Give two ways in which you could monitor the rate of reaction of calcium carbonate and dilute hydrochloric acid by experiment. [2 marks]

7 Ⓗ How can you find the rate of a reaction from a graph of mass of product against time? [2 marks]

8 Explain in terms of particles why increasing the concentration of a reactant increases the rate of the reaction. [2 marks]

9 Explain why increasing the temperature increases the rate of a reaction. [3 marks]

10 Ⓗ What effect will removing chlorine gas have on the equilibrium mixture shown in the equation:

$ICl(l) + Cl_2(g) \rightleftharpoons ICl_3(s)$ [2 marks]

11 Ⓗ Explain the effect of increasing the pressure on the equilibrium mixture shown in the equation:

$2SO_2(g) + O_2(g) \rightleftharpoons 2SO_3(g)$ [3 marks]

12 Ⓗ The reaction $CaCO_3(s) \rightleftharpoons CaO(s) + CO_2(g)$ reaches equilibrium in a closed system. The forward reaction is endothermic. How could the amount of calcium oxide produced by the reaction be increased? [1 mark]

Chapter checklist

Tick when you have:

reviewed it after your lesson ✓ ☐ ☐

revised once – some questions right ✓ ✓ ☐

revised twice – all questions right ✓ ✓ ✓

Move on to another topic when you have all three ticks

8.1 Rate of reaction ☐ ☐ ☐

8.2 Collision theory and surface area ☐ ☐ ☐

8.3 Effect of temperature ☐ ☐ ☐

8.4 The effect of concentration and pressure ☐ ☐ ☐

8.5 The effect of catalysts ☐ ☐ ☐

8.6 Reversible reactions ☐ ☐ ☐

8.7 Energy and reversible reactions ☐ ☐ ☐

8.8 Dynamic equilibrium ☐ ☐ ☐

8.9 Altering conditions ☐ ☐ ☐

Student Book
pages 148–149

C9

9.1 Hydrocarbons

Key points

- Crude oil is a mixture of many different compounds.

- Most of the compounds in crude oil are hydrocarbons – they contain only hydrogen and carbon atoms.

- Alkanes are saturated hydrocarbons. They contain as many hydrogen atoms as possible in their molecules.

- The general formula of an alkane is: $C_nH_{(2n+2)}$

Synoptic link

For more information about the Earth's finite (non-renewable) resources, see Topic C14.1.

Synoptic link

For information about covalent bonding, see Topic C3.5.

Key words: mixture, distillation, fraction, hydrocarbon, alkane, saturated hydrocarbon

Study tip

Draw a table with three columns: name, molecular formula, displayed formula. Complete the table for the first four alkanes.

- The large number of carbon compounds that occur is due to the ability of carbon atoms to form families of similar compounds. Carbon atoms can bond to each other to form chains and rings that form the 'skeletons' of molecules. These compounds are used as fuels to run cars, to warm homes, and to generate electricity.

- Crude oil is a finite resource found in rocks. It was formed over millions of years from the remains of tiny, ancient sea animals and plants, mainly plankton, that were buried in mud.

- Crude oil is a **mixture** of many different compounds that boil at different temperatures. The crude oil mixture is separated to make useful fuels.

- You can separate a mixture of liquids by **distillation**. Simple distillation of crude oil can produce liquids that boil within different temperature ranges. These liquid mixtures are called **fractions**.

1   What are fractions?

- Most of the compounds in crude oil are **hydrocarbons**. This means that their molecules contain only hydrogen and carbon. Many of these hydrocarbons are **alkanes**, with the general formula $C_nH_{(2n+2)}$. You describe alkanes as **saturated hydrocarbons**. All the carbon–carbon bonds are single covalent bonds. This means that they contain as many hydrogen atoms as possible in each molecule. No more hydrogen atoms can be added.

- The formulae of the first four alkane molecules are:
CH_4 (methane)
C_2H_6 (ethane)
C_3H_8 (propane)
C_4H_{10} (butane)

2   How can you tell that the substance with the formula C_9H_{20} is an alkane?

- You can represent molecules in different ways. A molecular formula shows the number of each type of atom in each molecule, for example, C_2H_6 represents a molecule of ethane. You can also represent molecules by a displayed formula that shows how the atoms are bonded together.

You can represent alkanes like this, showing all of the atoms and bonds in the molecule. These are called displayed formulae. The line drawn between two atoms in a molecule represents a single covalent bond

3   What is the molecular formula of butane?

9.2 Fractional distillation of oil

Key points

- Crude oil is separated into fractions in industry using fractional distillation.
- The properties of each fraction depend on the size of the hydrocarbon molecules in it.
- Lighter fractions make better fuels as they ignite more easily and burn well, with cleaner (less smoky) flames.

Synoptic link

For information about the process of fractional distillation, see Topic C1.4.

Key words: fractional distillation, viscosity, flammable

- Crude oil is separated into fractions at refineries using **fractional distillation**. This can be done because the boiling point of a hydrocarbon depends on the size of its molecule. The larger the molecule, the higher the boiling point of the hydrocarbon.

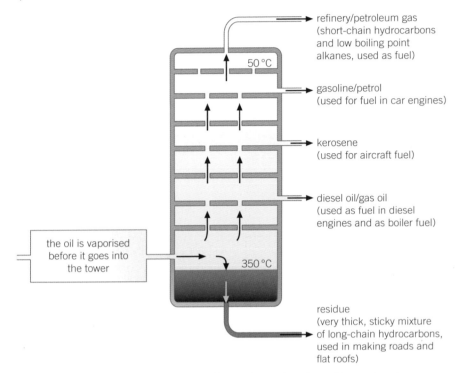

The boiling point of a hydrocarbon depends on the size of its molecules. Differences in boiling points can be used in fractional distillation to separate the hydrocarbons in crude oil into fractions

- The crude oil is vaporised and fed into a fractionating column. This is a tall tower that is hot at the bottom and gets cooler going up the column.
- Inside the column there are many trays with holes to allow gases through. The vapours move up the column getting cooler as they go up. The hydrocarbons condense to liquids when they reach the level that is at their boiling point. Different liquids collect on the trays at different levels and there are outlets to collect the fractions.
- Hydrocarbons with the smallest molecules have the lowest boiling points, so these gases are collected at the top of the column. The fractions collected at the bottom of the column contain hydrocarbons with the highest boiling points. These are solids at room temperature.

On the up

You should be able to make predictions about the properties of crude oil fractions from the length of the fraction's hydrocarbon chain.

1   Why are different hydrocarbons collected at different levels of a fractional distillation column?

- Fractions with smaller hydrocarbon molecules and lower boiling temperatures have lower **viscosities** so they are runny liquids. They are very **flammable** so they ignite easily. They also burn with clean flames, producing little smoke. This makes them very useful as fuels.

9.3 Burning hydrocarbon fuels

Key points

- When hydrocarbon fuels are burnt in plenty of air, the carbon and hydrogen in the fuel are completely oxidised. They produce carbon dioxide and water.

- You can test the gases formed in complete combustion of a hydrocarbon: the carbon dioxide turns limewater cloudy, and the water turns blue cobalt chloride paper pink (or white anhydrous copper sulfate blue).

- Incomplete combustion of a hydrocarbon produces carbon monoxide (a toxic gas) as one of its products.

Synoptic links

Testing for gases is covered in more detail in Topic C12.3.

To find out more about the effects of the pollution caused by burning hydrocarbon fuels, see Topic C13.5.

Key word: oxidised

- Burning any fuel that contains carbon produces carbon dioxide. When pure hydrocarbons burn completely they are **oxidised** to carbon dioxide and water.

  For example, when propane burns:

  propane + oxygen → carbon dioxide + water

  $$C_3H_8 + 5O_2 \rightarrow 3CO_2 + 4H_2O$$

1 Write a word equation for the complete combustion of ethane.

- The products given off when a hydrocarbon burns can be tested as shown below.

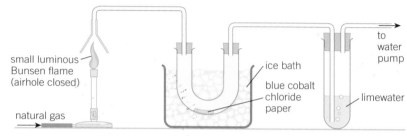

small luminous Bunsen flame (airhole closed)

natural gas

ice bath

blue cobalt chloride paper

limewater

to water pump

Testing the products of complete combustion

- To test for water you use blue cobalt chloride paper. Blue cobalt chloride paper turns pink in contact with water.

- Limewater turns cloudy when carbon dioxide is present.

- Sometimes the fuels you use are not burnt completely. In a limited supply of air, incomplete combustion may produce a toxic gas, carbon monoxide (CO).

On a winter's day you can often see the water produced when the hydrocarbons in petrol or diesel burn, as the steam formed in combustion of the fuel condenses when it cools down

2 Name two possible products of the incomplete combustion of a hydrocarbon.

9.4 Cracking hydrocarbons

- Large hydrocarbon molecules can be broken down into smaller molecules by a process called **cracking**.

- Cracking can be done in two ways:

  - steam cracking, by heating a mixture of hydrocarbon vapours and steam to a very high temperature

  - catalytic cracking, by passing hydrocarbon vapours over a hot catalyst.

- During cracking, **thermal decomposition** reactions produce a mixture of smaller molecules. Some of the smaller molecules are alkanes, which are saturated hydrocarbons with the general formula C_nH_{2n+2}. These alkanes with smaller molecules are more useful as fuels.

1   Give one reason why an oil company might want to crack large hydrocarbons to make smaller alkanes.

In an oil refinery, huge crackers like this are used to break down large hydrocarbon molecules into smaller, more useful ones. The petrochemical industry source the chemicals used to make products such as solvents, lubricants, polymers, and detergents, as well as the fuels that modern life depends on

- Some of the other smaller molecules formed in cracking are hydrocarbons with the general formula C_nH_{2n}. These are called **alkenes**. Alkenes are **unsaturated hydrocarbons** because they contain fewer hydrogen atoms than alkanes with the same number of carbon atoms. Alkenes have a C=C double covalent bond in their molecules.

- This is an example of a cracking reaction.

$$C_{10}H_{22} \xrightarrow{\text{500 °C + catalyst}} C_5H_{12} + C_3H_6 + C_2H_4$$
$$\text{decane} \qquad\qquad \text{pentane} \quad \text{propene} \quad \text{ethene}$$

ethene — double bond

propene

A molecule of ethene, C_2H_4, and a molecule of propene, C_3H_6. These are both alkenes – each molecule has a carbon–carbon double bond in it

- Alkenes are used to produce polymers and other chemicals such as solvents. Alkenes have a **double bond** between two carbon atoms and this makes them more reactive than alkanes. Alkenes react with bromine water turning it from orange to colourless. This reaction is used as a test for unsaturated compounds.

2   Give three ways in which alkenes are different from alkanes.

Key points

- Large hydrocarbon molecules can be broken down into smaller molecules by passing the vapours over a hot catalyst, or by mixing them with steam and heating them to a very high temperature.

- Cracking produces saturated hydrocarbons, used as fuels, and unsaturated hydrocarbons (called alkenes).

- Alkenes (and other unsaturated compounds containing carbon–carbon double bonds) react with orange bromine water, turning it colourless.

Synoptic link

To find out more about the reaction between an alkene and bromine water, see Topic C10.1.

Key words: cracking, thermal decomposition, alkene, unsaturated hydrocarbon, double bond

Study tip

Different mixtures of alkanes and alkenes can be produced by cracking because different hydrocarbons can be used and the conditions for the reaction can be changed.

On the up

You should be able to describe the process of cracking, including conditions, and complete balanced equations.

1 Why is crude oil separated into fractions? [1 mark]

2 Name the products when ethane, C_2H_6, burns completely, that is, undergoes complete combustion. [2 marks]

3 Why are some scientists concerned about the carbon dioxide produced by burning fossil fuels? [2 marks]

4 Give two reasons why fractions from crude oil are cracked. [2 marks]

5 Describe two ways that are used to crack hydrocarbons. [2 marks]

6 Copy and complete this equation for cracking a hydrocarbon:
$C_{12}H_{26} \rightarrow C_6H_{14} + C_4H_8 +$ _____ [1 mark]

7 Give three reasons why fractions with lower boiling points are more useful as fuels. [3 marks]

8 Propane, C_3H_8, is used as a fuel for cookers. Explain why propane should always be burnt in a plentiful supply of air. [3 marks]

9 Pentane has the formula C_5H_{12}.
  a Draw a displayed formula for pentane. [1 mark]
  b Write down four facts about pentane that you can deduce from its formula. [4 marks]

10 How many carbon atoms are there in an alkane molecule with 40 hydrogen atoms? [1 mark]

11 Write a balanced symbol equation for the complete combustion of butane. [2 marks]

12 Explain what happens in a fractional distillation column used to separate crude oil. [5 marks]

Chapter checklist

Tick when you have:

reviewed it after your lesson	✓	☐	☐
revised once – some questions right	✓	✓	☐
revised twice – all questions right	✓	✓	✓

Move on to another topic when you have all three ticks

9.1 Hydrocarbons	☐	☐	☐
9.2 Fractional distillation of oil	☐	☐	☐
9.3 Burning hydrocarbon fuels	☐	☐	☐
9.4 Cracking hydrocarbons	☐	☐	☐

10.1 Reactions of the alkenes

Key points

- The general formula of the alkenes, containing one C=C bond, is C_nH_{2n}
- Complete combustion of an alkene forms carbon dioxide and water.
- Alkenes react with halogens, hydrogen, and water (steam) by adding atoms across the C=C bond, forming a saturated molecule.

An alkene will decolourise orange bromine water because the dibromoethane formed is colourless

Study tip

Draw the displayed formula of propene. Annotate the diagram to describe the features of the molecule.

Synoptic link

For more information on chemical equilibrium, see Topic C8.8.

On the up

You should be able to compare and contrast the reactivity of alkanes and alkenes.

- The alkenes all have a C=C group of atoms in their molecules. The C=C grouping is an example of a **functional group**. A 'family' of organic compounds with the same functional group is called a **homologous series**.

ethene, C_2H_4

propene, C_3H_6

butene, C_4H_8

pentene, C_5H_{10}

The first four members of the homologous series of alkenes. The lines between atoms represent covalent bonds. Notice the carbon–carbon double bond (C=C) in each alkene molecule.

- Alkenes are unsaturated hydrocarbons because they contain two fewer hydrogen atoms than alkanes with the same number of carbon atoms.

- The general formula for alkenes is: C_nH_{2n}
  The formulae of the first four alkene molecules are:

  C_2H_4 (ethene) C_4H_8 (butene)
  C_3H_6 (propene) C_5H_{10} (pentene)

1 What is the formula for an alkene containing six carbon atoms?

- Complete combustion of alkenes gives carbon dioxide and water. For example:
  ethene + oxygen → carbon dioxide + water
  $$C_2H_4 + 3O_2 \rightarrow 2CO_2 + 2H_2O$$
  However, alkenes tend to burn in air with a smokier flame than alkanes because incomplete combustion takes place.

Addition reactions of the alkenes

- The carbon–carbon double bond, C=C, makes the alkenes more reactive than the alkanes. Alkenes react by addition reactions. Atoms are added across the carbon–carbon double bond, so that a carbon–carbon single bond is formed.

- Alkenes react with halogens. The halogen atoms add across the double bond. For example, ethene reacts with bromine to produce dibromoethane.

- Alkenes react with hydrogen to produce alkanes. The reaction takes place at 60 °C in the presence of a nickel catalyst. For example:

  catalyst
  butene + hydrogen → butane
  $$C_4H_8 + H_2 \rightarrow C_4H_{10}$$

- Alkenes react with steam to produce alcohols. For example, ethene gas can react with water (steam) in the presence of a catalyst to produce ethanol. The reaction is reversible.

  catalyst
  ethene + steam ⇌ ethanol
  $$C_2H_4 + H_2O \rightleftharpoons C_2H_5OH$$

Key words: functional group, homologous series

10.2 Structures of alcohols, carboxylic acids, and esters

Key points

- The homologous series of alcohols contains the –OH functional group.
- The homologous series of carboxylic acids contains the –COOH functional group.
- The homologous series of esters contains the –COO– functional group.

- Alcohols contain the functional group –OH. If one hydrogen atom from each alkane molecule is replaced with an –OH group, you get a homologous series of alcohols.
- The first four members of this series are methanol, ethanol, propanol, and butanol.
- A structural formula shows which atoms are bonded to each carbon atom and the functional group. The structural formula of ethanol is CH_3CH_2OH.

methanol   ethanol   propanol   butanol

The displayed formulae of the first four members of the alcohol homologous series.

1   Write the structural formula of propanol.

- Carboxylic acids have the functional group –COOH.
- The first four members of the carboxylic acids are methanoic acid, ethanoic acid, propanoic acid, and butanoic acid. Their structural formulae are HCOOH, CH_3COOH, CH_3CH_2COOH, and $CH_3CH_2CH_2COOH$.

methanoic acid   ethanoic acid   propanoic acid   butanoic acid

The displayed formulae of the first four carboxylic acids

Study tip

If you draw displayed formulae, make sure you show all the bonds, including those in the functional group (as lines between atoms) and all the atoms (as their chemical symbols).

2   Draw the displayed formula of methanoic acid.

- Esters have the functional group –COO–. If the H atom in the –COOH group of a carboxylic acid is replaced by a hydrocarbon group the compound is an ester.
- Ethyl ethanoate has the structural formula $CH_3COOCH_2CH_3$.

ethyl ethanoate

The displayed formula of the ester ethyl ethanoate

10.3 Reactions and uses of alcohols

Key points

- Alcohols are used as solvents and fuels, and ethanol is the main alcohol in alcoholic drinks.
- Alcohols burn in air, forming carbon dioxide and water.
- Alcohols react with sodium metal to form a solution of sodium alkoxide, and hydrogen gas is given off.
- Ethanol can be oxidised to ethanoic acid, either by chemical oxidising agents or by the action of microbes in the air. Ethanoic acid is the main acid in vinegar.

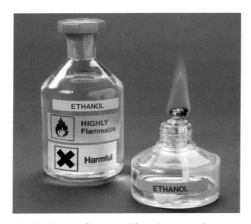

Alcohols are flammable. They produce carbon dioxide and water in their combustion reactions

Key word: fermentation

Study tip

You should be able to identify the products of reactions of alcohols but you only need to be able to write balanced equations for the combustion reactions of alcohols.

- Alcohols with smaller molecules, such as methanol, ethanol, and propanol, mix well with water and produce neutral solutions.
- Many organic substances dissolve in alcohols and so they are used as solvents in products such as perfumes and mouthwashes.
- Ethanol is the main alcohol in wine, beer, and other alcoholic drinks. Ethanol is manufactured by **fermentation**, in which yeast (a fungus) converts sugars (from plant material) into an aqueous solution of ethanol.
- In some countries, ethanol is available as an alternative fuel to petrol or diesel. Ethanol made by fermentation is termed a biofuel.

1 Why do many perfumes contain ethanol?

- Alcohols burn in air. When burnt completely, alcohols produce carbon dioxide and water. Alcohols are used as fuels and they can be mixed with petrol.
- Ethanol is used as a biofuel in cars:
  ethanol + oxygen → carbon dioxide + water
  $$C_2H_5OH + 3O_2 \rightarrow 2CO_2 + 3H_2O$$
- Sodium reacts with alcohols to produce hydrogen gas, but the reaction is less vigorous than when sodium reacts with water.
- Alcohols can be oxidised by chemical oxidising agents such as potassium dichromate to produce carboxylic acids. Some microbes in the air can also oxidise solutions of ethanol to produce ethanoic acid, which turns alcoholic drinks sour and is the main acid in vinegar.

2 Ethanol and water are both colourless liquids. Suggest one chemical test you could do to tell them apart.

10.4 Carboxylic acids and esters

Key points

- Solutions of carboxylic acids have a pH value of less than 7. Carbonates gently fizz in solutions of carboxylic acids, releasing carbon dioxide gas.

- Aqueous solutions of carboxylic acids, which are weak acids, have a higher pH value than solutions of strong acids of the same concentration.

- Esters are made by reacting a carboxylic acid with an alcohol, in the presence of a strong acid catalyst.

- Esters are volatile, fragrant compounds used in flavourings and perfumes.

Carboxylic acids produce carbon dioxide gas when a metal carbonate is added

Synoptic link

For more information on the differences between strong and weak acids, see Topic C5.8.

- Carboxylic acids dissolve in water to produce solutions with a pH value of less than 7. They have the properties typical of all acids. For example, when carboxylic acids are added to carbonates they fizz because they react to produce carbon dioxide. A salt and water are also produced.

1   Why do carboxylic acids have properties similar to all other acids?

Synoptic link

See Chapter C5 for the typical reactions of acids.

- Carboxylic acids are different from other acids because they react with alcohols in the presence of an acid catalyst to produce esters. For example, ethanol and ethanoic acid react together when mixed with sulfuric acid as a catalyst, to produce ethyl ethanoate and water:

$$\text{ethanoic acid} + \text{ethanol} \underset{}{\overset{\text{sulfuric acid catalyst}}{\rightleftharpoons}} \text{ethyl ethanoate} + \text{water}$$

In general:

$$\text{carboxylic acid} + \text{alcohol} \underset{}{\overset{\text{strong acid catalyst}}{\rightleftharpoons}} \text{ester} + \text{water}$$

- Esters are volatile compounds and have distinctive smells. Some esters have pleasant fruity smells and are used as flavourings and in perfumes.

2   Why are some esters used as flavourings?

Carboxylic acids are weak acids. Carbon dioxide is given off more slowly when carboxylic acids react with a metal carbonate compared with hydrochloric acid of the same concentration. This is because carboxylic acids are weak acids. When ethanoic acid dissolves in water, it does not ionise completely and most of the ethanoic acid molecules remain as molecules in the solution:

$$CH_3COOH(aq) \rightleftharpoons CH_3COO^-(aq) + H^+(aq)$$

Higher

3   Write a balanced equation to show that propanoic acid is a weak acid.

Hydrochloric acid is a strong acid that ionises completely in aqueous solutions to form hydrogen ions and chloride ions.

$$HCl(aq) \rightarrow H^+(aq) + Cl^-(aq)$$

On the up

To achieve the highest grades, you should be able to explain why carboxylic acids are weak acids and know how to tell the difference between strong and weak acids.

1. Which formulae represent alkenes?
   C_3H_6 C_5H_{12} C_4H_{10} C_4H_8 C_6H_{14} C_8H_{16} [3 marks]

2. Why is ethene more reactive than ethane? [1 mark]

3. Name the gas produced when ethanoic acid is added to sodium carbonate. [1 mark]

4. Name and give the structural formula of the carboxylic acid with three carbon atoms in its molecule. [2 marks]

5. Name and give the structural formulae of the first three members of the series of alcohols. [6 marks]

6. Describe what happens when a small piece of sodium is added to some ethanol in a beaker. [2 marks]

7. A glass of beer containing 5% ethanol was left exposed to the air for 12 hours. The beer turned sour. Explain why. [3 marks]

8. Ethanol and ethanoic acid can react together to produce an ester.
   a Name the ester. [1 mark]
   b Describe the conditions used for the reaction. [2 marks]

9. Describe how ethanol is produced from:
   a glucose [2 marks]
   b ethene. [2 marks]

10. Why are alkenes not used as fuels? [3 marks]

11. Write a balanced symbol equation for the complete combustion of propanol. [2 marks]

12. You have been given aqueous solutions of hydrochloric acid and ethanoic acid that have the same concentration. Suggest one simple test that you could do to decide which solution is ethanoic acid. [2 marks]

13. **H** Explain why hydrochloric acid is described as a strong acid whereas ethanoic acid is described as a weak acid. [2 marks]

Chapter checklist

Tick when you have:

reviewed it after your lesson ✓ ☐ ☐
revised once – some questions right ✓ ✓ ☐
revised twice – all questions right ✓ ✓ ✓
Move on to another topic when you have all three ticks

10.1 Reactions of the alkenes ☐ ☐ ☐

10.2 Structures of alcohols, carboxylic acids, and esters ☐ ☐ ☐

10.3 Reactions and uses of alcohols ☐ ☐ ☐

10.4 Carboxylic acids and esters ☐ ☐ ☐

11.1 Addition polymerisation

Key points

- Plastics are made of very large, covalently bonded molecules called polymers.
- The large polymer molecules are made when many monomers (small, reactive molecules) join together.
- The reaction between alkene monomers to form a polymer is called addition polymerisation.
- In addition polymers, the repeating unit has the same atoms as the monomer, because when the C=C bond 'opens up' in polymerisation, no other molecule is formed in the reaction.

Polymers produced from compounds derived from crude oil are all around you, and are part of your everyday life

Synoptic links

The products obtained from crude oil are described in Topics C9.2 and C9.4.

Key words: polymer, monomer, addition polymerisation

- Plastics are made of very large molecules called **polymers**. Polymers are made from many small molecules joined together. The small molecules used to make polymers are called **monomers**. The reaction to make a polymer is called polymerisation.

1 How are polymers made?

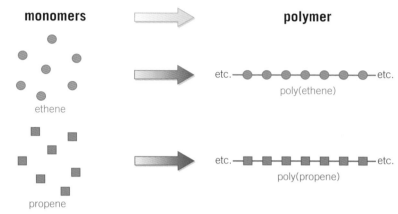

Polymers are made from many smaller molecules called monomers

- In **addition polymerisation**, an addition polymer is formed when thousands of alkene molecules join together. The double covalent bond between two carbon atoms in each molecule 'opens up' and becomes a single carbon–carbon covalent bond between the two carbon atoms. A polymer chain is formed. Only one product is formed in the reaction.

- You can show what happens when ethene (C_2H_4) monomers join together to form the addition polymer poly(ethene):

ethene monomers → poly(ethene)

- You can write this more simply as a chemical equation:

$$ n \quad \underset{\substack{\text{many single}\\\text{ethene monomers}}}{\text{C}=\text{C}} \longrightarrow \underset{\substack{\text{long chain}\\\text{of poly(ethene)}}}{\left(\!\!\left[\text{C}-\text{C}\right]\!\!\right)_n} $$

where *n* is a large number

- In addition polymers, the repeating unit of the polymer has the same atoms as the monomer. The repeating unit in poly(ethene) is shown bracketed in the chemical equation above.

2 How many monomers are there in a poly(ethene) molecule?

- Other alkenes can polymerise in a similar way. For example, propene (C_3H_6), can form poly(propene).

3 Why can we make polymers from alkenes but not from alkanes?

11.2 Condensation polymerisation

Key points

- Condensation polymerisation usually involves a small molecule released in the reaction, as the polymer forms.

- The monomers used to make the simplest condensation polymers are usually two different monomers, with two of the same functional groups on each monomer.

- Polyesters are formed from the condensation polymerisation of a diol and a dicarboxylic acid, with H_2O given off in the reaction.

This cotton/polyester fabric is durable and does not crease as easily as pure cotton fabric

The difference between addition and condensation polymers

- In addition polymerisation, only one product is formed. In **condensation polymerisation**, there are two different products. The main product is the polymer, but you also get a small molecule, usually water, given off.

    addition polymerisation → the addition polymer

    condensation polymerisation → the condensation polymer + a small molecule

- Examples of condensation polymers are polyester and nylon.

> 1 How many products are formed in condensation polymerisation?

- In addition polymerisation the monomers are usually the same alkene, containing C=C bonds. In condensation polymerisation there are often two different monomers used. Each monomer has a functional group at both ends of its molecule. The functional groups on the two different monomers react together to form long polymer chains.

Forming a polyester

- A polyester is made from a diol (an alcohol containing two —OH groups) and a dicarboxylic acid (a carboxylic acid containing two —COOH groups). The monomers link together and polymerise by 'ester links'. A water molecule is given off as each link is made in the reaction:

$$n\,HO-CH_2-CH_2-OH \;+\; n\,HOOC-CH_2-CH_2-CH_2-CH_2-COOH$$

a diol   a dicarboxylic acid

(ethanediol)   (hexanedioic acid)

↓

$$\left(\!CH_2-CH_2-OOC-CH_2-CH_2-CH_2-CH_2-COO\!\right)_n \;+\; 2nH_2O$$

a polyester   water

or generally represented as:

$$n\,HO-\square-OH + n\,HOOC-\square-COOH \longrightarrow \left(\!\square-OOC-\square-COO\!\right)_n + 2nH_2O$$

> 2 Which functional groups join together to form a polyester?

Synoptic link

For an explanation of how an ester is made, see Topic C10.4.

Key word: condensation polymerisation

Study tip

Draw a table to compare addition and condensation polymers.

On the up

You should be able to compare and contrast, giving appropriate examples, addition polymerisation and condensation polymerisation.

11.3 Natural polymers

Key points

- Simple carbohydrates (monosaccharides) polymerise to make polymers such as starch and cellulose.

- Proteins are polymers made from different amino acid monomers.

- **H** Amino acids have an acidic and a basic functional group in the same molecule.

- **H** Amino acids react together during condensation polymerisation to make polypeptides and proteins made of long sequences of different monomers.

Key word: protein

- There are naturally occurring polymers such as starch, cellulose, and **proteins** found in living things.

- The most commonly known sugar is glucose, $C_6H_{12}O_6$. Glucose is a monosaccharide (made of one sugar unit). Monosaccharide sugars can act as monomers to make polymers, called polysaccharides.

  glucose monomers → starch polymers + water

  glucose monomers → cellulose polymers + water

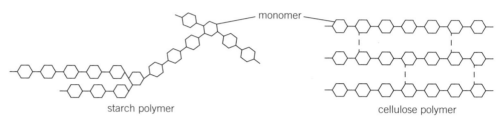

  starch polymer   cellulose polymer

The diagram shows simplified structures of starch and cellulose, both are made from glucose monomers joined in condensation polymerisation

1   Name two polymers formed from glucose.

Proteins are also natural polymers, formed by condensation polymerisation.

  variety of amino acid monomers → protein polymers + water

The monomers of proteins are called amino acids. Amino acids have two functional groups in a molecule: the basic amine group, $—NH_2$, and the acidic carboxylic acid group, $—COOH$. The simplest amino acid is glycine, H_2NCH_2COOH.

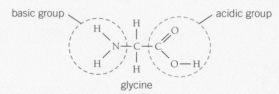

  basic group   acidic group

  glycine

The displayed formula of glycine. Its structural formula is H_2NCH_2COOH

Many glycine molecules link together in a condensation reaction to form a polypeptide molecule. You can show this as:

$$nH_2NCH_2COOH \xrightarrow{\text{polymerisation}} -(HNCH_2CO)_n- + nH_2O$$

2   Why is the formation of a protein classed as condensation polymerisation?

- There are about 20 different amino acids that join together in many different sequences to make the proteins in your body.

Higher

11.4 DNA

Key points

- DNA (deoxyribonucleic acid) is made up from monomers called nucleotides.

- The nucleotides are based on the sugar deoxyribose, bonded to a phosphate group and a base. There are four different nucleotides.

- A DNA molecule consists of two polymer strands (with sugars bonded to phosphate groups) intertwined into a double helix.

Key words: DNA (deoxyribonucleic acid), nucleotide

Study tip

Write general equations to show how starch, cellulose, proteins, and DNA are formed.

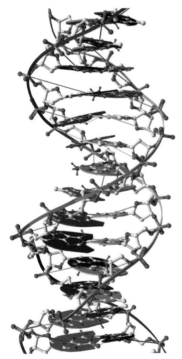

The double helix of DNA

- **DNA (deoxyribonucleic acid)** is a natural polymer that is essential for life. DNA's structure contains a genetic code that allows living organisms and viruses to develop and function.

- DNA is made by the condensation polymerisation of repeating units of monomers called **nucleotides**. So DNA is known as a polynucleotide:

  millions of nucleotides → DNA (a polynucleotide) + water

1 What is the general name of the monomers used to make DNA?

- The DNA molecule consists of a double helix made up of two long polymer strands (chains). The two strands run in opposite directions to each other. DNA is made from four different nucleotides.

- The four different nucleotides consist of amino acids attached to a sugar called deoxyribose and a phosphate group. The nucleotides join by forming covalent bonds between the sugars of one nucleotide and the phosphate of another. The strands are held in place by the intermolecular forces down the length of each polymer strand.

On the up

You should be able to describe and explain the shape of the DNA polymer.

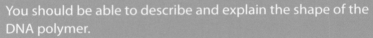
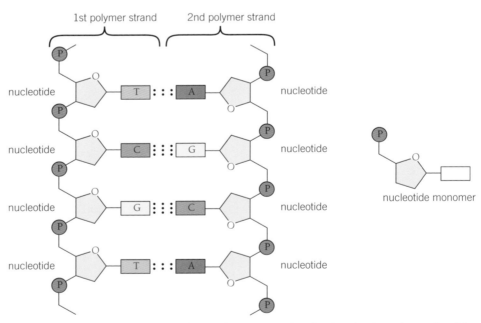

A simplified (flattened) structure of part of the DNA molecule, showing how the nucleotides bond to each other in each strand (via covalent bonds) and how the strands are linked to each other by intermolecular forces

2 How many different nucleotides are found in DNA?

1　State two uses of poly(ethene). [2 marks]

2　**a** Name the shape of a DNA molecule. [1 mark]
　　b How many polymer chains are found in a DNA molecule? [1 mark]

3　**a** Name the monomer that poly(propene) is produced from. [1 mark]
　　b Give the chemical formula of this monomer. [1 mark]

4　**a** What is the difference between natural and synthetic polymers? [2 marks]
　　b Name three natural polymers and the monomers they are made from. [3 marks]

5　**a** How many different types of monomer are found in DNA? [1 mark]
　　b Why is DNA called a polynucleotide? [2 marks]

6　Ⓗ How does the number of products differ in the reactions to form an addition polymer and a condensation polymer? [2 marks]

7　Ⓗ Write the general equation for the reaction between a diol and a dicarboxylic acid. [1 mark]

8　Poly(ethene) is a polymer.
　　Explain what is meant by a polymer. [4 marks]

9　Ⓗ Describe two differences between the monomers used to make an addition polymer and a condensation polymer. [4 marks]

10　Write an equation for the polymerisation of ethene using displayed formulae showing the bonds. [2 marks]

11　Polymers made from different monomers have different properties.
　　Explain why. [2 marks]

12　Ⓗ Explain how amino acids combine to form a protein. [5 marks]

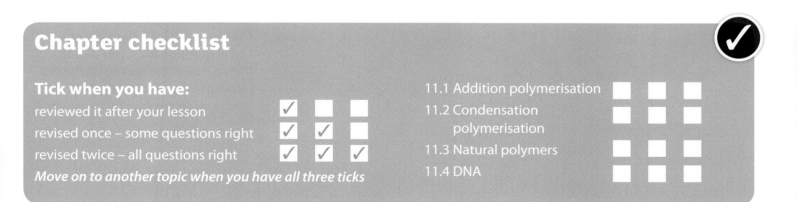

Chapter checklist

Tick when you have:

reviewed it after your lesson ✔ ☐ ☐
revised once – some questions right ✔ ✔ ☐
revised twice – all questions right ✔ ✔ ✔
Move on to another topic when you have all three ticks

11.1 Addition polymerisation ☐ ☐ ☐
11.2 Condensation polymerisation ☐ ☐ ☐
11.3 Natural polymers ☐ ☐ ☐
11.4 DNA ☐ ☐ ☐

01 The chemical formula of glucose is $C_6H_{12}O_6$.
01.1 Name the elements in glucose. [2 marks]
01.2 How many atoms are present in one molecule of glucose? [1 mark]
01.3 Name **two** naturally occurring polymers made from glucose. [2 marks]

02 Methanol, ethanol, and propanol are the first three members of a homologous series of compounds.
02.1 What general name is used for members of this homologous series? [1 mark]
02.2 Write the molecular formula of propanol. [1 mark]
02.3 What is the functional group that all members of this homologous series have? [1 mark]

03 The equation shows the reaction to form white anhydrous copper sulfate crystals by heating blue hydrated copper sulfate.
$$CuSO_4.5H_2O \rightleftharpoons CuSO_4 + 5H_2O$$
      blue   white
03.1 What type of reaction does the symbol $\rightleftharpoons$ represent? [1 mark]
03.2 To produce anhydrous copper sulfate, hydrated copper sulfate is heated. What type of reaction is this? [1 mark]
03.3 How could you change the white anhydrous copper sulfate into blue hydrated copper sulfate? [1 mark]
03.4 The reaction in **03.3** is an exothermic reaction. What energy change will be occurring? [1 mark]
03.5 What temperature change would you observe when the reaction in **03.3** is done? [1 mark]
03.6 A student was trying to identify a colourless liquid. The student thought the liquid was water.
Suggest how the student could use this reaction to test for water. [1 mark]
03.7 Describe what the student in **03.6** would observe if the substance was water. [1 mark]

04 The structure of ethene is shown in **Figure 1**.

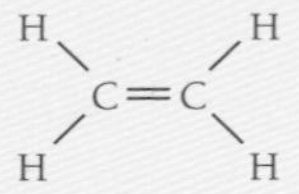

ethene,
C_2H_4

Figure 1

04.1 Ethene is a hydrocarbon. What is meant by a hydrocarbon? [2 marks]
04.2 Name the functional group present in ethene. [1 mark]
04.3 Describe a chemical test to show the presence of the group in **04.2**. [2 marks]
04.4 Copy and complete the equation for the complete combustion of ethene. [3 marks]

$$C_2H_4 + 3O_2 \rightarrow \underline{\qquad\qquad} + \underline{\qquad\qquad}$$

04.5 Ethene can burn with a smoky flame. Explain why. [2 marks]

> **Study tip**
>
> When asked for a chemical test, as in question **04.3**, you should always describe how to do the test and give the result of the test.

04.6 The equation shows how ethene can form a polymer.

$$n \; \underset{\underset{H}{\overset{H}{|}}}{\overset{\overset{H}{|}}{C}} = \underset{\underset{H}{\overset{H}{|}}}{\overset{\overset{H}{|}}{C}} \; \longrightarrow \; \left(\underset{\underset{H}{\overset{H}{|}}}{\overset{\overset{H}{|}}{C}} - \underset{\underset{H}{\overset{H}{|}}}{\overset{\overset{H}{|}}{C}} \right)_n$$

What is the name of the polymer formed? [1 mark]

04.7 Describe what is happening to the ethene molecules in **04.6** during the reaction. [3 marks]

05 A group of students investigated the decomposition of hydrogen peroxide to produce oxygen gas and water.
The equation for the reaction is:

$$2H_2O_2(aq) \rightarrow 2H_2O(l) + O_2(g)$$

The reaction is catalysed by manganese(IV) oxide.
The students added 2 g of manganese(IV) oxide to 50 cm^3 of hydrogen peroxide solution at 20 °C. They measured the volume of oxygen produced.
Table 1 shows their results.

Table 1

Time in seconds	0	20	40	60	80	100	120	140	160	180
Volume of oxygen in cm^3	0	18	34	48	59	65	74	78	80	80

05.1 Plot a graph of the students' results. Draw a line of best fit through the points. [5 marks]

05.2 Identify the anomalous point. [1 mark]

05.3 How can you tell from the graph that the rate was fastest at the start of the reaction? [1 mark]

05.4 Explain in terms of particles why the rate of reaction was fastest at the beginning. [2 marks]

05.5 Ⓗ How could the students use the graph to find the rate of reaction after 40 seconds? [2 marks]

05.6 The students repeated the investigation at 30 °C. All other variables were kept the same.
Sketch a line on your graph to show the results you would expect for this experiment. [2 marks]

05.7 Suggest **two** control variables the students used. [2 marks]

06 Oil companies crack some of the fractions from crude oil. The equation shows an example of a reaction that happens during cracking.

$$C_\_H_\_ \rightarrow C_7H_{16} + C_4H_8 + C_3H_6$$

06.1 Copy and complete the molecular formula of the reactant. [1 mark]

06.2 What conditions are used for cracking? [2 marks]

06.3 Draw a diagram to show all of the bonds in C_3H_6. [3 marks]

06.4 C_3H_6 can be used to make a polymer.
Which other product of the reaction above can be used to make a polymer? [1 mark]

06.5 Plastic waste that contains polymers made from C_3H_6 and similar monomers causes environmental problems. Explain why. [6 marks]

Study tip

When you are asked for a specific number of answers, as in question **05.7**, you should only give the number of answers that are required. If you give additional incorrect answers you will lose marks.

Study tip

Some longer questions, such as question **06.5**, may have a level of response mark scheme. In these mark schemes, there is not a mark for each scientific response that you make. It is the overall quality of the answer that you write that decides the mark.

4 Analysis and the Earth's resources

Analytical chemists have developed many tests to detect specific chemicals. These tests are based on reactions that produce a gas with distinctive properties, or a colour change or an insoluble solid that appears as a precipitate. Instrumental analysis provides fast, sensitive, and accurate results, as used by forensic scientists and anti-doping scientists.

The Earth's atmosphere is dynamic and always changing. Some of these changes are man-made and some are part of natural cycles. Scientists and engineers are trying to solve the problems caused by increased levels of air pollutants. Industries use the Earth's natural resources to manufacture useful products. In order to operate sustainably, chemists seek to minimise the use of limited resources, the use of energy, waste produced, and environmental impact.

I already know...

I will revise...

I already know...	I will revise...
about the difference between pure substances and mixtures and how to identify some pure substances.	a wider range of chemical tests to identify unknown gases and ions and why instrumental analysis is used in many applications.
the composition of the atmosphere.	how the atmosphere developed over the Earth's history before arriving at its current composition.
the production of carbon dioxide by human activity and its impact on climate.	how climate change is caused by increasing levels of greenhouse gases and how this issue needs to be addressed.
about the Earth as a source of limited resources and the efficacy of recycling.	how to analyse data on our diminishing finite resources, including order of magnitude estimations, and carry out Life Cycle Assessments to judge the impact of making new materials.
some properties of ceramics, polymers, and composites.	how to explain the properties of ceramics, polymers and composites in terms of their chemical structures.
the use of carbon in obtaining metals from metal oxides.	the use of biological methods to extract some metals, such as copper, from low grade deposits of metal ores.

12.1 Pure substances and mixtures

Key points

- Pure substances can be compounds or elements, but they contain only one substance. An impure substance is a mixture of two or more different elements or compounds.

- Pure elements and compounds melt and boil at specific temperatures, and these fixed points can be used to identify them.

- Melting point and boiling point data can be used to distinguish pure substances (specific fixed points) from mixtures (that melt or boil over a range of temperatures).

- Formulations are useful mixtures, made up in definite proportions, designed to give a product the best properties possible to carry out its function.

- In everyday life, pure can mean 'has had nothing added to it', or that it is in its natural state. For example, pure orange juice means from freshly squeezed oranges.

- In chemistry, a pure substance is one that is made up of just one element or compound. The element or compound is not mixed with any other substance.

- A test for water is that it turns white anhydrous copper sulfate blue. The test only shows that water is present, not that the water is pure. The test for pure water is that its melting point is exactly 0 °C, and its boiling point is exactly 100 °C.

- You can use melting points or boiling points to identify pure substances because pure substances have specific temperatures at which they melt and boil. The melting and boiling points of an element or a compound are called its fixed points. They can be looked up in databases.

1   In chemistry, what is a pure substance?

- A mixture does not have a sharp melting point or boiling point. It changes state over a range of temperatures.

- You can find a melting point (or melting range) to distinguish between pure and impure substances. Impurities lower the melting point of a substance and raise its boiling point.

- A **formulation** is a mixture that has been designed to produce a useful product. Each chemical in the mixture has a specific purpose. The chemicals are mixed in set proportions to give the product the required properties.

- An example of a formulation is paint. Paints contain:

  - a pigment, to provide colour

  - a binder, to help the paint attach to an object

  - a solvent, to help the pigment and binder spread well during painting.

- Formulations include fuels, alloys, fertilisers, cleaning agents, paints, medicines, cosmetics, and food products.

Synoptic link

To find out about the differences between mixtures and compounds, see Topic C1.3.

Synoptic links

Nanoparticles are becoming an important component in many formulations, such as cosmetics and paints. To find out more about nanoparticles, see Topics C3.11 and C3.12.

Key word: formulation

2   A substance boils between 98 °C and 102 °C. Is this a pure substance or a mixture?

Study tip

Write a tweet (maximum 140 characters) to describe each of the following terms: pure substance, impure substance, formulation.

On the up

You should be able to justify the classification of pure substances, impure substances, and formulations when data is supplied.

12.2 Analysing chromatograms

Key points

- Scientists can analyse unknown substances in solution by using paper chromatography.

- R_f values can be measured and matched against databases to identify specific substances.

- $R_f = \dfrac{\text{distance moved by substance}}{\text{distance moved by solvent}}$

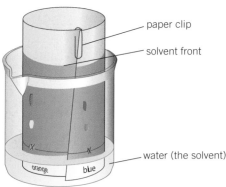

A chromatogram is the paper record of the separation

Key word: R_f (retention factor)

On the up

You should be able to calculate R_f values when given data or from a chromatogram.

- In paper chromatography the mobile phase is the solvent and the stationary phase is the paper. The solvent moves through the paper carrying different substances different distances. A substance with stronger forces of attraction to the solvent than the paper will travel further in a set time.

- A mixture of compounds will probably form more than one spot on the chromatogram. A pure substance will only form a single spot in all solvents.

1 What are the two phases in chromatography?

Detecting dyes in food colourings

You can use paper chromatography to analyse various food colourings. Draw a pencil line near the bottom of a sheet of chromatography paper. Use a capillary tube to dab a spot of the solution on the pencil line.

Stand the paper in the solvent at the bottom of the beaker. The pencil line should be above the solvent level to start with.

Allow the solvent to travel up the paper, removing the paper before the solvent reaches the top.

Mark the solvent front on the chromatography paper.

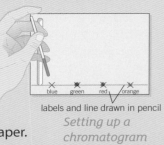

Setting up a chromatogram

- The more soluble a substance is in the solvent, the further up the paper it is carried.

- You can compare the spots on the chromatogram with others obtained from known substances. The data is presented as an R_f **(retention factor)** value.

$$R_f = \frac{\text{distance moved by substance}}{\text{distance moved by solvent}}$$

Worked example

Find the R_f value of compounds A and B using the chromatogram below.

The R_f value of A $= \dfrac{\text{distance moved by substance}}{\text{distance moved by solvent}}$

$= \dfrac{8}{12} = \mathbf{0.67}$

The R_f value of B $= \dfrac{3}{12} = \mathbf{0.25}$

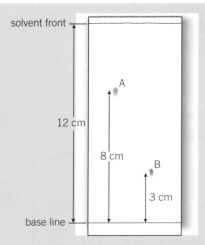

The R_f value of an unknown substance, in a particular solvent at a given temperature, can be compared with values in a database to identify the substance

Student Book
pages 184–185 **C12** # 12.3 Testing for gases

Key points

- Hydrogen gas burns rapidly with a 'pop' when you apply a lighted splint.
- Oxygen gas relights a glowing splint.
- Carbon dioxide gas turns limewater milky (cloudy).
- Chlorine gas bleaches damp blue litmus paper white.

- Hydrogen gas: Collect a test tube of gas. Hold a lighted splint at the open end of the test tube. Positive test for hydrogen: the hydrogen burns rapidly with a 'squeaky pop' sound.

- Oxygen gas: Insert a glowing splint (made by blowing out a lighted splint) in the mouth of the test tube. Positive test for oxygen gas: the glowing splint relights.

- Carbon dioxide gas: Bubble the gas through limewater (calcium hydroxide solution). Positive test for carbon dioxide gas: limewater turns milky (cloudy white).

- Chlorine: Chlorine is a toxic gas, so care must be taken when working with this gas. A piece of damp blue litmus paper is held in the mouth of the test tube. Positive test for chlorine gas: damp blue litmus paper is bleached and turns white.

1 Which gas relights a glowing splint?

Student Book
pages 186–187 **C12** # 12.4 Tests for positive ions

Key points

- Some metal ions (including most Group 1 and 2 cations) can be identified in their compounds using flame tests.
- Sodium hydroxide solution can be used to identify metal ions that form insoluble hydroxides in precipitation reactions.

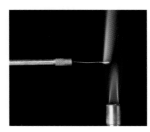

A flame test can identify some metal ions in compounds

On the up

You should be able to write balanced symbol equations for the reactions in these tests. To achieve the highest grades, you should also be able to write the ionic equations.

- Some positive ions can be identified using a flame test or by using sodium hydroxide solution.

- Some metal ions produce colours when put into a Bunsen flame. Lithium (Li^+) = crimson; sodium (Na^+) = yellow; potassium (K^+) = lilac; calcium (Ca^{2+}) = orange-red; copper (Cu^{2+}) = green.

- The hydroxides of most metals that have ions with 2+ and 3+ charges are insoluble in water. When you add sodium hydroxide solution to solutions of these ions, a precipitate of the metal hydroxide forms.

- Aluminium, calcium, and magnesium ions all form white precipitates of their hydroxides. When you add excess sodium hydroxide solution only the white precipitate of aluminium hydroxide dissolves, showing aluminium ions are present.

- Some metal ions form coloured precipitates with sodium hydroxide. If you add sodium hydroxide solution to a substance containing:
  - copper(II) ions, a blue precipitate of copper(II) hydroxide appears
  - iron(II) ions, a green precipitate of iron(II) hydroxide is produced
  - iron(III) ions, a brown precipitate of iron(III) hydroxide is formed.

- You can show the reactions of solutions of metal ions with sodium hydroxide solution by a balanced symbol equation. For example, if you test copper(II) sulfate solution with sodium hydroxide solution, the precipitation reaction is:

$$CuSO_4(aq) + 2NaOH(aq) \longrightarrow Na_2SO_4(aq) + Cu(OH)_2(s)$$

The ionic equation for the reaction is:

H

$$Cu^{2+}(aq) + 2OH^-(aq) \longrightarrow Cu(OH)_2(s)$$

1 A few drops of sodium hydroxide solution were added to a colourless solution and a white precipitate appeared. When excess sodium hydroxide was added the precipitate remained. Which metal ions could be present?

12.5 Tests for negative ions

Key points

- You identify carbonates by adding dilute acid, which produces carbon dioxide gas. The gas turns limewater milky (cloudy).
- You identify halides by adding nitric acid, then silver nitrate solution.
- This produces a precipitate of silver halide (chloride = white, bromide = cream, iodide = yellow).
- You identify sulfates by adding hydrochloric acid, then barium chloride solution. This produces a white precipitate of barium sulfate.

Study tip

In the test for sulfate ions, add hydrochloric acid before the barium chloride. Do not add sulfuric acid – it contains sulfate ions!

- Carbonate ions: Add dilute hydrochloric acid to the substance to see if it fizzes. If it does and the gas produced turns limewater cloudy, the substance contains carbonate ions.

1   Which gas turns limewater cloudy?

- Halide ions: Add dilute nitric acid and then silver nitrate solution.
  - Chloride ions give a white precipitate.
  - Bromide ions give a cream precipitate.
  - Iodide ions give a yellow precipitate.

  For example, $NaCl(aq) + AgNO_3(aq) \longrightarrow NaNO_3(aq) + AgCl(s)$

- Sulfate ions: Add dilute hydrochloric acid and then barium chloride solution. If a white precipitate forms, sulfate ions are present.

  If the unknown compound is sodium sulfate, then the equation for the precipitation reaction is: $Na_2SO_4(aq) + BaCl_2(aq) \longrightarrow 2NaCl(aq) + BaSO_4(s)$

2   Why must you add nitric acid and not hydrochloric acid or sulfuric acid when testing with silver nitrate solution for halides?

Identifying unknown ionic compounds

You should be able to identify the ions present in an unknown compound using the tests for positive and negative ions.

12.6 Instrumental analysis

Key points

- Modern instrumental techniques provide fast, accurate, and sensitive ways of analysing chemical substances.
- Flame emission spectroscopy is an example of an instrumental method.
- This method indicates which metal ions are present from their characteristic line spectra, and also gives the concentration of the metal ions in a solution.

- Modern instrumental methods of analysis are rapid, accurate, and sensitive, often using very small samples. Computers process the data from the instruments to give results almost instantly. The equipment is usually very expensive and needs special training to use it.

1   Give two advantages of using an instrumental method of analysis.

- **Flame emission spectroscopy** is an instrumental method that chemists use to analyse samples for metal ions. The sample is heated in a flame. You analyse the light produced by passing it through a spectroscope. Each type of metal ion has its own characteristic pattern, called its line spectrum. You compare the line spectra produced with a database to identify the metal ions present. The concentration of the metal ions present in a sample can also be determined.

2   How do you use the line spectrum to identify a substance in flame emission spectroscopy?

Key word: flame emission spectroscopy

1. What substances in foods can be detected by paper chromatography? [1 mark]

2. What colour do copper ions give in a flame test? [1 mark]

3. What is the positive test for the presence of sulfate ions in a compound? [3 marks]

4. Why is 'pure orange juice' an incorrect statement in chemistry? [2 marks]

5. Explain why paint is a formulation. [3 marks]

6. Give two advantages and two disadvantages of using modern instrumental techniques of analysis. [4 marks]

7. a Name the instrumental technique scientists use to identify the metal ions in a mixture of substances. [1 mark]

   b Why don't scientists use flame tests to identify the metal ions? [3 marks]

8. On a paper chromatogram, a spot was obtained from an unknown substance 5 cm above the base line. The solvent front was measured at 20 cm above the base line. What is the R_f value of this substance? [2 marks]

9. Dilute hydrochloric acid was added to a green compound. The mixture gave off a gas that turned limewater cloudy and a blue solution was formed. When sodium hydroxide solution was added to the blue solution a blue precipitate was produced. Name the green compound. [2 marks]

10. A compound gave a lilac colour in a flame test. Nitric acid and silver nitrate solution were added to a solution of the compound and a yellow precipitate was formed. Name the compound. [2 marks]

11. When sodium hydroxide solution was added to a solution a green precipitate formed. When hydrochloric acid and barium chloride solution were added to another sample of the solution a white precipitate formed. Which ions were in the solution? [2 marks]

12. Ⓗ Write an ionic equation to show aqueous iron(III) ions reacting with hydroxide ions in solution. Include state symbols in your equation. [3 marks]

Chapter checklist ✓

Tick when you have:

reviewed it after your lesson	✓	☐	☐
revised once – some questions right	✓	✓	☐
revised twice – all questions right	✓	✓	✓

Move on to another topic when you have all three ticks

12.1 Pure substances and mixtures	☐	☐	☐
12.2 Analysing chromatograms	☐	☐	☐
12.3 Testing for gases	☐	☐	☐
12.4 Tests for positive ions	☐	☐	☐
12.5 Tests for negative ions	☐	☐	☐
12.6 Instrumental analysis	☐	☐	☐

13.1 History of our atmosphere

Key points

- The Earth's early atmosphere was formed by volcanic activity.
- It probably consisted mainly of carbon dioxide. There may also have been nitrogen and water vapour, together with traces of methane and ammonia.
- As plants spread over the Earth, the levels of oxygen in the atmosphere increased.

Key word: atmosphere

- Scientists think that the Earth was formed about 4.6 billion years ago.
- There are several theories about how Earth's early **atmosphere** formed, but there is little direct evidence from billions of years ago.
- One theory suggests the Earth's surface was covered in volcanoes. In the first billion years, intense volcanic activity released the gases that formed the early atmosphere.

Volcanoes moved chemicals from inside the Earth to the surface and the newly forming atmosphere

- Earth's early atmosphere was probably mainly carbon dioxide. There could also have been some water vapour, nitrogen gas, and traces of methane and ammonia. There would have been very little or no oxygen at that time. This resembles the atmospheres today on the planets Mars and Venus.
- As the Earth cooled down, water vapour in the atmosphere condensed to form oceans.

1 Where did the gases in the early atmosphere come from?

- Scientists think that life began about 3.4 billion years ago. Algae evolved about 2.7 billion years ago. Algae used energy transferred as light waves from the Sun to photosynthesise. More and more plants evolved. The plants photosynthesised, removing carbon dioxide from, and releasing oxygen into, the atmosphere.

$$\text{carbon dioxide} + \text{water} \xrightarrow{\text{light}} \text{glucose} + \text{oxygen}$$

$$6CO_2 + 6H_2O \rightarrow C_6H_{12}O_6 + 6O_2$$

- Over the next billion years, the levels of oxygen rose steadily. This made it possible for the first animals to evolve.

2 What process produced the oxygen in the atmosphere?

13.2 Our evolving atmosphere

Key points

- Photosynthesis by algae and plants decreased the percentage of carbon dioxide in the early atmosphere. The formation of sedimentary rocks and fossil fuels that contain carbon also removed carbon dioxide from the atmosphere.

- Any ammonia and methane was removed by reactions with oxygen, once oxygen had been formed by photosynthesis.

- Approximately four-fifths (about 80%) of the atmosphere today is nitrogen, and about one-fifth (about 20%) is oxygen.

- There are also small proportions of various other gases, including carbon dioxide, water vapour, and noble gases.

There is clear fossil evidence in carbonate rocks of the organisms which lived millions of years ago

Study tip

Draw a spider diagram to show how carbon dioxide was removed from Earth's early atmosphere.

- Scientists think Earth's early atmosphere was mainly carbon dioxide, but now the percentage in the atmosphere is only about 0.04%.

- Algae and plants decreased the percentage of carbon dioxide in the early atmosphere by photosynthesis.

- Carbon dioxide gas was also removed from the early atmosphere by dissolving in oceans. Insoluble carbonate compounds were produced. These then precipitated on the seabed as sediments.

- Over millions of years, animal remains became covered with layers of sediment at the bottom of the oceans. Under the pressure of the sediment, the deposits formed sedimentary carbonate rocks. An example is limestone, a rock consisting mainly of calcium carbonate, formed from the shells and skeletons of marine organisms.

- The fossil fuels coal, crude oil, and natural gas were formed from plant and animal remains.

  - Coal is a sedimentary rock, and was formed from thick deposits of plant material. The plant material was buried, in the absence of oxygen and compressed over millions of years.

  - Crude oil and natural gas were formed from the remains of plankton deposited in mud on the seabed. These remains were covered by sediments that became layers of rock when compressed over millions of years. The crude oil and natural gas were trapped in the layers of rock.

1   Give three ways in which carbon dioxide was removed from the early atmosphere.

- Volcanoes also produced nitrogen gas, which gradually built up in the early atmosphere, and there may have also been small proportions of methane and ammonia gases. The methane and ammonia were removed by reacting with the oxygen produced by photosynthesis. These reactions produced carbon dioxide and nitrogen.

- However, the levels of nitrogen gas in the atmosphere built up because nitrogen is a very unreactive gas.

- Over the last 200 million years, the proportions of gases in the Earth's atmosphere have not changed much.

- The atmosphere is now almost four-fifths (approximately 80%) nitrogen and just over one-fifth (approximately 20%) oxygen.

- Other gases in the atmosphere include carbon dioxide, water vapour, and noble gases. Together these make up about 1% of the atmosphere.

2   What are the approximate percentages of nitrogen and oxygen in the air?

13.3 Greenhouse gases

- Greenhouse gases in the atmosphere maintain temperatures on Earth high enough to support life. The main greenhouse gases in the Earth's atmosphere are carbon dioxide, methane, and water vapour.

1 Name three greenhouse gases.

- The Earth is heated by the Sun. The greenhouse gases let short-wavelength radiation pass through the atmosphere to the Earth's surface.

- The surface of the Earth cools down by emitting longer wavelength infrared radiation. However, greenhouse gases absorb infrared radiation. So some of the energy radiated from the surface of the Earth gets trapped in the atmosphere and the temperature rises. The higher the proportion of greenhouse gases in the air, the more energy is absorbed.

- Over the past 100 years, the amount of carbon dioxide released into the atmosphere has greatly increased. The main reason for this is the increased use of fossil fuels, which when burnt release carbon dioxide into the atmosphere.

- In deforestation, trees are cut down for timber and to clear land. This means less carbon dioxide is removed from the air because photosynthesis is reduced.

- As the temperature rises, carbon dioxide gets less soluble in water. This makes the oceans less effective as 'CO$_2$ sinks'.

2 What is the main reason for the increase in carbon dioxide in the Earth's atmosphere?

- The increasing amount of methane in the air is due to:
  - more animal farming leading to more emissions and decomposing waste
  - increased waste, so more decomposition of rubbish in landfill sites.

- Most scientists agree that a trend in global warming has started. Their views are based on evidence presented in scientific journals. This evidence is peer reviewed by other scientists.

- Scientists are searching for further evidence of the link between the levels of greenhouse gases and the climate. However, even using the most powerful computers, it is difficult to model such complex systems as global climate change. Therefore you cannot predict with certainty the effects of increasing levels of greenhouse gases. This can lead to simplified models or opinions based on only parts of the evidence, which may be biased.

Key points

- The amount of carbon dioxide in the Earth's atmosphere has risen in the recent past, largely due to the amount of fossil fuels now burnt.

- It is difficult to predict with complete certainty the effects on climate of rising levels of greenhouse gases on a global scale.

- However, the vast majority of peer-reviewed evidence agrees that increased proportions of greenhouse gases from human activities will increase average global temperatures.

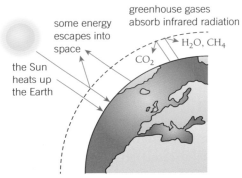

carbon dioxide, methane, and water vapour are the main greenhouse gases

The molecules of a 'greenhouse gas' absorb the energy radiated by the Earth as it cools down at night. This increases the store of energy of the gases in the atmosphere and warms the Earth

On the up

You should be able to explain how human activities change the proportion of greenhouse gases in the atmosphere.

13.4 Global climate change

Key points

- Reducing greenhouse gases in the atmosphere relies on reducing the use of fossil fuels, mainly by using alternative sources of energy and conserving energy.

- The economies of developed countries are based on energy obtained from fossil fuels, so changes will cost money to implement.

- However, changes are needed because of the potential risks arising from global climate changes, such as rising sea levels, threats to ecosystems, and different patterns of food production around the world.

Hurricanes could become more common in some areas as a consequence of global climate change

Synoptic link

To find out about assessing the life cycle of a product in terms of its environmental impact, see Topic C14.5.

Key words: climate change, carbon footprint, biofuel, carbon capture and storage

- Global warming is a major cause of **climate change**.

- The consequences of rising levels of greenhouse gases are:

  - rising sea levels, as a result of melting ice caps, causing flooding and increased coastal erosion

  - more frequent and severe storms

  - changes in temperature and the amount and distribution of rainfall

  - impact on the food-producing capacity of different regions.

  - changes to the distribution of wildlife species, with some becoming extinct.

1   What is the main cause of climate change?

- The **carbon footprint** of a product, service, or event is the total amount of carbon dioxide and other greenhouse gases emitted over its full life cycle.

- Actions to reduce the carbon footprint include:

  - increased use of alternative energy supplies

  - conserving energy, for example, by improving home insulation

  - using 'carbon neutral' fuels. **Biofuels** made from plant material absorb carbon dioxide during photosynthesis, and return this to the atmosphere when they are burnt

  - carbon taxes – taxing fossil fuels, for example, or cars that burn a lot of petrol or diesel

  - ability to offset carbon taxes by planting trees – an incentive for companies that produce CO_2 in their processing of materials

  - **carbon capture and storage** – carbon dioxide produced in fossil fuel power stations is pumped deep underground and absorbed by porous rocks. However, the cost of producing electricity will be increased by about 10%

  - plant-based diets, because farmers grow more crops and vegetables, and fewer cattle produce less methane.

- Problems in reducing the carbon footprint include:

  - scientific disagreement over the causes and consequences of global climate change

  - incomplete international co-operation on setting targets for the reduction of greenhouse gas emissions

  - cost implications in all manufacturing and transport industries

  - restrictions on greenhouse gases could hinder the developing industries of poorer countries

  - lack of public information and education, for example, encouraging people to make lifestyle changes such as walking, cycling, and recycling.

2   Why are biofuels said to be 'carbon neutral'?

13.5 Atmospheric pollutants

Key points

- When hydrocarbon fuels are burnt in plenty of air, the carbon and hydrogen in the fuel are completely oxidised. They produce carbon dioxide (the main greenhouse gas) and water.

- Sulfur impurities in fuels burn to form sulfur dioxide, which can cause acid rain. Sulfur can be removed from fuels before they are burnt, or sulfur dioxide can be removed from flue gas.

- Changing the conditions in which hydrocarbon fuels are burnt can change the products made.

- In insufficient oxygen, poisonous carbon monoxide gas is formed. Particulates of carbon (soot) and unburnt hydrocarbons can also be produced, especially if the fuel is diesel. They can cause global dimming.

- At the high temperatures in engines, nitrogen from the air reacts with oxygen to form oxides of nitrogen. These cause breathing problems and can also cause acid rain.

Study tip

Draw a spider diagram to show how atmospheric pollutants are produced from the combustion of fossil fuels.

On the up

You should be able to evaluate the scale, risk, and environmental impact of global climate change.

- When pure hydrocarbons burn completely they are oxidised to carbon dioxide and water. For example, when butane burns:

  butane + oxygen → carbon dioxide + water
  $$2C_4H_{10} + 13O_2 \rightarrow 8CO_2 + 10H_2O$$

- However, the fuels you use do not always burn completely and may also contain other substances. In a limited supply of air, **incomplete combustion** may produce carbon monoxide, some unburnt hydrocarbons, and **particulates** (solid particles) that contain soot (carbon).

- Carbon monoxide is a toxic gas. It is colourless and odourless, so you cannot tell that you are breathing it in. The carbon monoxide combines with haemoglobin in the blood and is carried around in your blood instead of oxygen.

- Particulates can cause global dimming as they prevent some sunlight reaching the Earth's surface. Particulates also cause lung damage in humans.

1   Name four possible products of the incomplete combustion of a hydrocarbon.

- Most fossil fuels contain sulfur compounds. When the fuel burns, these sulfur compounds produce sulfur dioxide (SO_2).

- At the high temperatures produced when fuels burn, oxygen and nitrogen in the air may combine to form **nitrogen oxides**.

- Sulfur dioxide and nitrogen oxides cause acid rain, as well as breathing problems in humans. Acid rain damages trees, as well as killing animal and plant life in lakes. Acid rain also attacks limestone buildings and metal structures.

2   What environmental problem is caused by sulfur dioxide and nitrogen oxides?

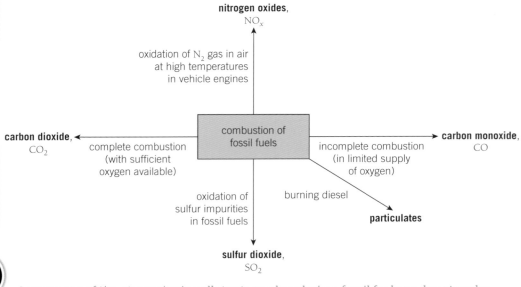

A summary of the atmospheric pollutants produced when fossil fuels are burnt under different conditions

Key words: incomplete combustion, particulate, nitrogen oxides

1 Name four gases in Earth's atmosphere today. [4 marks]

2 What was the main gas in Earth's early atmosphere? [1 mark]

3 Name and give the formulae of three greenhouse gases. [6 marks]

4 Name the process by which oxygen was produced in the early atmosphere. [1 mark]

5 What type of radiation do greenhouse gases absorb? [1 mark]

6 State two human activities that release greenhouse gases into the air. [2 marks]

7 List three practical ways you can make a positive contribution to reducing emissions of greenhouse gases. [3 marks]

8 **a** List four possible effects of increasing levels of greenhouse gases. [4 marks]

   b Why is it difficult to predict the long-term effects of increasing levels of greenhouse gases? [2 marks]

9 Why is it important to have a plentiful supply of air when burning methane? [3 marks]

10 What is the name and chemical formula of the main constituent of limestone? [2 marks]

11 Explain why removing sulfur impurities from fossil fuels before they are burnt has a positive effect on plant and animal life living in lakes. [5 marks]

12 Write the word and symbol equation for the reaction in which oxygen and glucose are produced in plants. [3 marks]

13 Write a balanced symbol equation for the complete combustion of propane, C_3H_8. [2 marks]

14 Explain how limestone is formed from the shells and skeletons of marine animals. [5 marks]

Chapter checklist

Tick when you have:

reviewed it after your lesson ✓ ☐ ☐

revised once – some questions right ✓ ✓ ☐

revised twice – all questions right ✓ ✓ ✓

Move on to another topic when you have all three ticks

13.1 History of our atmosphere ☐ ☐ ☐

13.2 Our evolving atmosphere ☐ ☐ ☐

13.3 Greenhouse gases ☐ ☐ ☐

13.4 Global climate change ☐ ☐ ☐

13.5 Atmospheric pollutants ☐ ☐ ☐

14.1 Finite and renewable resources

Key points

- We rely on the Earth's natural resources to make new products and provide us with energy.

- Some of these natural resources are finite – they will run out eventually if we continue to exploit them, for example, fossil fuels.

- Others are renewable – they can be replaced as we use them up, for example, crops used to make biofuels.

- Estimates of the time left before fossil fuels run out can only be rough estimates, because of the uncertainty involved in the calculations.

Synoptic link

For more information about poly(ethene) and how it is made, see Topic C11.1.

On the up

You should be able to interpret information and draw conclusions about resources from graphs, charts, and tables.

- Humans rely on the Earth's natural resources to live. Natural resources are used to provide food, warmth, shelter, and transport. Natural resources are found in the Earth's crust, oceans, and atmosphere.

- Farming of plants and animals has increased the supply of some of these products. Chemists have also developed synthetic alternatives to some natural resources.

Natural resource	Use	Alterative synthetic product
wool	clothes, carpets	acrylic fibre, poly(propene)
cotton	clothes, textiles	polyester, nylon
rubber	tyres	synthetic polymers
wood	construction	PVC, composites

- You can classify natural resources as finite or renewable.

- Finite resources are being used up at a faster rate than they can be replaced. Finite resources will eventually run out, for example, fossil fuels.

- Renewable resources are those that can be replaced at the same rate at which they are used up, for example, the crops used to make biofuels.

1   Name two natural resources.

- The chemical industry uses natural resources as the raw materials to make new products, for example:

  - metal ores used to extract metals

  - crude oil used to make polymers and petrochemicals

  - limestone to make cement and concrete

  - crude oil to make petrol and diesel.

- Industries are moving towards renewable resources to conserve finite resources and to improve sustainability. Sustainability is developments that meet the needs of society now, without endangering the ability of future generations to meet their needs. Two examples of sustainability are:

  - In the plastics industry, polymers are produced from ethene. Ethene can be made from crude oil or ethanol. Ethanol can be produced from sugar cane. Using a renewable crop as the raw material for ethene makes the poly(ethene) produced more sustainable.

  - The use of wood chips instead of fossil fuels to fuel power stations. Planting new trees will provide a source of wood chips in the future.

2   Why is using wood chips instead of fossil fuels to fuel a power station more sustainable?

14.2 Water safe to drink

Key points

- Water is made fit to drink by passing it through filter beds to remove solids and adding chlorine, ozone, or by passing ultra-violet light through it (sterilising) to reduce microorganisms.

- Water can be purified by distillation, but this requires large amounts of energy, which makes it expensive.

- Reverse osmosis uses membranes to separate dissolved salts from salty water, but this method of desalination also needs large amounts of energy to create the high pressures needed.

- Water is a vital resource and is essential for life, agriculture, and industry.

- Drinking water should not contain any harmful substances and should have sufficiently low levels of dissolved salts and microorganisms. Water that is safe to drink is called potable water. Potable water is not pure water because it contains dissolved substances.

- Rainwater collects in rivers and lakes and is called fresh water. Fresh water contains dissolved minerals (salts).

- When water is taken from an appropriate source, it has to be treated to make it potable. The water is treated by passing it through filter beds to remove solid particles. The water is sterilised (microorganisms are killed) by adding chlorine or ozone, or by passing ultraviolet light through the water.

1   How is water treated to make it fit to drink?

- In the UK, there are sufficient supplies of fresh water. Countries with few sources of fresh water may use other methods to obtain potable water.

- Seawater or salty water can be made potable by desalination. However, desalination is an expensive process, because of the large energy costs involved in heating the salty water. Pure water can be made by distillation.

- A process called reverse osmosis is also used to desalinate water. This uses membranes to separate the water from the salts dissolved in it.

About 97% of the water on Earth is found in its oceans, which is not potable water

Study tip

Drinking or potable water is not pure water, but it should not contain anything that will cause you harm.

Analysis and purification of water samples

To test the pH of a sample you can use universal indicator or a pH meter.

You can test for pure water by measuring its boiling point. Pure water boils at 100 °C.

To distil salt water, you would use the apparatus below.

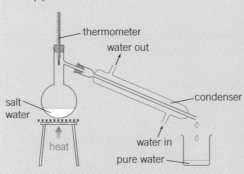

Distilling pure water from salt water

To test for dissolved solids, you would use the apparatus below.

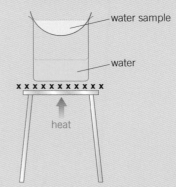

Crystallising dissolved solids in a water sample

2   How can you test for pure water?

14.3 Treating waste water

Key points

- Waste water requires treatment at a sewage works before being released into the environment.
- Sewage treatment involves the removal of organic matter and harmful microorganisms and chemicals.
- The stages include screening to remove large solids and grit, sedimentation to produce sewage sludge, and aerobic biological treatment of the safe effluent released into environment.
- The sewage sludge is separated, broken down by anaerobic digestion and dried. It can provide us with fertiliser and a source of renewable energy.

- Sewage is the general name for waste water from homes, businesses, and industry. This, together with waste water from farming activities, has to be treated at sewage treatment plants.

- Sewage treatment removes organic matter and harmful microorganisms from sewage and agricultural waste water. Industrial waste water may also need harmful chemicals removing. This process makes sewage safe so it can be returned to the environment.

1   What is sewage?

- Sewage treatment involves a series of steps.
  - Screening – this removes large solid objects and grit from the waste water.
  - Sedimentation – solid sediments (sludge) settle out from the mixture. The watery liquid above the sludge is called effluent. Effluent still contains many potentially harmful microorganisms.
  - Aerobic biological treatment – in the effluent useful bacteria feed on any remaining organic matter and harmful microorganisms. This breaks them down aerobically (in the presence of oxygen).

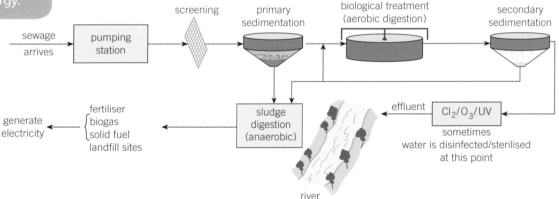

The steps needed to make our waste water from urban and rural sources safe to return to the environment

- The sludge contains organic matter, including human waste, suspended solids, water, and dissolved compounds. It is digested anaerobically (without oxygen) by microorganisms. The dried sewage sludge can then be used as a fertiliser or as a source of renewable energy.

2   Why are useful bacteria added to sludge?

14.4 Extracting metals from ores

Key points

- Most copper is extracted by smelting (roasting) copper-rich ores, although supplies of ores are becoming scarcer.

- Copper can be extracted from solutions of copper compounds by electrolysis or by displacement using scrap iron. Electrolysis is also used to purify impure copper, for example, the copper metal obtained from smelting.

- Scientists are developing ways to extract copper that use low-grade copper ores. Bacteria are used in bioleaching and plants in phytomining.

Synoptic link

For information on the displacement (redox) reaction between copper ions in solution and iron metal, see Topic C5.2.

Key word: bioleaching

- Most copper is extracted from copper-rich ores. There is a limited supply of this finite resource.

- Copper is produced from copper-rich ores by smelting. Smelting and purifying copper ore uses huge amounts of energy. This is very expensive and causes environmental pollution.

- Scientists are developing new techniques using bacteria (**bioleaching**) or plants (phytomining) to extract copper from low-grade ores. These biological methods have less environmental impact than smelting.

> 1 Give one advantage of using biological methods rather than smelting to extract copper.

- Phytomining grows plants on soil containing low-grade copper ore. The plants absorb copper ions. Then the plants are burnt and copper is extracted from copper compounds in the ash.

- Bioleaching uses bacteria to produce a solution of copper compounds (leachate) from waste copper ore.

- Scrap iron can then be added to solutions of the copper compounds to displace the copper.

- Electrolysis can also be used to extract pure copper from solutions of copper compounds.

> 2 Why does scrap iron displace copper from a solution of a copper compound?

Mining copper ores can leave huge scars on the landscape. Our supplies of copper-rich ores are a limited, finite resource

14.5 Life Cycle Assessments

Key points

- Life Cycle Assessments (LCAs) are carried out to assess the environmental impact of products, processes, or services.
- They analyse each of the stages of a life cycle, from extracting and processing raw material to disposal at the end of its useful life, including all transport and distribution.
- Data is available for the use of energy, water, resources, and production of some wastes.
- However, assigning numerical values to the relative effects of pollutants involves subjective judgements, so LCAs using this approach must make this uncertainty clear.

- **Life Cycle Assessments (LCA)** are used to assess the impact on the environment caused by manufacturing products.
- The four stages in an LCA can be summarised as:

  Raw material extraction → Manufacture → Use/Reuse/Maintenance → Recycle/Waste management

- To carry out an LCA you list all the inputs (energy and raw materials) and all the outputs (into the environment). Then evaluate the potential environmental impacts and interpret the results to help make decisions about using processes, products, or services.
- Numerical data can be used in LCAs for energy, water, resources, and production of some wastes. Allocating values to environmental impacts is more difficult. This requires subjective, not objective, scientific judgments to be made. It is hard to compare numerically global impacts (e.g., global warming), to local impacts (e.g., toxic emissions into a stream). Therefore the results of an LCA will be open to discussion.
- An LCA should be peer reviewed to check the data and validity of conclusions drawn. This is especially desirable if the LCA is devised by the company making the product being assessed, and its results are used to make claims in advertising.

1 Why should an LCA be peer reviewed?

Key word: Life Cycle Assessment (LCA)

14.6 Reduce, reuse, and recycle

Key points

- There are social, economic, and environmental issues associated with exploiting the Earth's limited supplies of raw materials, such as metal ores.
- Recycling metals saves energy and our limited, finite metal ores (and fossil fuels). The pollution caused by the mining and extraction of metals is also reduced by recycling.

Key words: non-renewable, recycle, blast furnace

- The aim of the 'reduce, reuse, and recycle' campaign is to reduce the use of limited resources and energy sources, produce less waste, and reduce pollution.
- Many products such as metals, glass, building materials, clay, ceramics, and most plastics are produced from limited supplies of raw materials. To convert raw materials into the finished products requires a lot of energy. Much of this energy is transferred from limited **non-renewable** resources, such as fossil fuels.
- Some products, such as glass bottles, can be reused. Glass can be crushed and melted to make different glass products. Other products cannot be reused, but can be **recycled** for a different use, for example, cars.
- Recycling metals conserves the limited reserves of metal ores. The processes to extract the metals from their ores have environmental impacts, such as acid rain and global warming. Mining and quarrying to obtain the metal ore can cause visual, noise, and dust pollution as well as destroying habitats. Large heaps of waste rock are also produced.
- Metals can be recycled by melting. Using recycled steel saves about 50% of the energy used to extract iron and produce steel. To conserve iron ore, some steel scrap is added to the **blast furnace**.

1 Why should we recycle aluminium cans?

1 **a** What is the difference between finite and renewable resources? [2 marks]

  b Classify the following as finite or renewable resources:
  sugar cane, crude oil, cotton, wood, coal, gold [6 marks]

2 Suggest three reasons why we should recycle metals. [3 marks]

3 Why is chlorine used in water treatment? [2 marks]

4 Why is sewage passed through a metal grid at the start of the sewage treatment process? [1 mark]

5 What is the difference between fresh water and potable water? [2 marks]

6 Why is bottled mineral water not pure water? [2 marks]

7 **a** What is sewage sludge? [2 marks]

  b Give two uses of sewage sludge. [2 marks]

8 **H** **a** Name two biological methods used to extract copper. [2 marks]

  b Give two advantages of using these methods. [2 marks]

9 **a** What method could you use to obtain pure water from seawater? [1 mark]

  b How could you test the water to see if it was pure? [2 marks]

  c What difference would you expect if you tested seawater in the same way? [1 mark]

10 **a** What is the purpose of an LCA? [1 mark]

  b What are the main stages in an LCA? [4 marks]

  c Why is it advisable to use an LCA from a different source rather than from the company making the product being assessed? [1 mark]

11 **a** To extract copper, copper sulfide (Cu_2S) can be roasted in air (oxygen) to produce copper and sulfur dioxide. Write a word equation for the process. [1 mark]

  b Write a balanced symbol equation for the process. [2 marks]

  c What environmental impact does sulfur dioxide cause? [1 mark]

12 **H** Copper can also be obtained by adding scrap iron to copper sulfate solution.

  a Write a balanced symbol equation with state symbols for the process. [3 marks]

  b Write an ionic equation for the process. [2 marks]

Chapter checklist

Tick when you have:

reviewed it after your lesson	✓	☐	☐
revised once – some questions right	✓	✓	☐
revised twice – all questions right	✓	✓	✓

Move on to another topic when you have all three ticks

14.1 Finite and renewable resources	☐	☐	☐
14.2 Water safe to drink	☐	☐	☐
14.3 Treating waste water	☐	☐	☐
14.4 Extracting metals from ores	☐	☐	☐
14.5 Life Cycle Assessments	☐	☐	☐
14.6 Reduce, reuse, and recycle	☐	☐	☐

Student Book
pages 220–221 **C15**

15.1 Rusting

- Corrosion is caused by chemical reactions between the metal and substances in the environment.

- The corrosion of iron is called **rusting**. Rust is a form of iron(III) oxide, called hydrated iron(III) oxide.

Key points

- Both air (oxygen) and water are needed for iron to rust.

- Providing a barrier between iron and any air (oxygen) and water protects the iron from rusting.

- Sacrificial protection provides protection against rusting, even when the iron is exposed to air and water. The iron needs to be attached to a more reactive metal (zinc, magnesium, or aluminium).

Investigating the conditions needed for iron to rust

Tube **A** tests to see if air alone will make iron rust.
Tube **B** tests to see if water alone will make iron rust.
Tube **C** tests to see if air and water will make iron rust.
The tubes were left for a week.

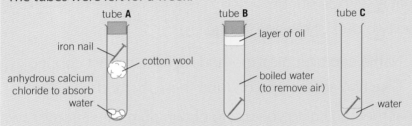

Investigating the conditions needed for iron to rust

After one week, the nail in tube **C** had rusted. The nails in tubes **A** and **B** were unchanged.

- The investigation shows that both air and water are needed for rusting to occur. The reaction to form rust can be shown as:

$$\text{iron} + \text{oxygen} + \text{water} \rightarrow \text{hydrated iron(III) oxide}$$

1   What conditions are needed for iron to rust?

- Coating the metal with paint, grease, or plastic can prevent corrosion. The coating forms a protective barrier around the metal. The iron is not in contact with the air and water needed for rusting.

- Some coatings use a more reactive metal to provide protection. Even if the coating is scratched, the iron does not rust.

- **Galvanised** iron is iron covered with a layer of zinc. Zinc is more reactive than iron. This is because zinc has a stronger tendency to form positive ions by giving away electrons. Therefore, any water or oxygen reacts with the zinc rather than the iron. This is called **sacrificial protection**. The zinc is sacrificed to protect the iron.

- Aluminium metal does not corrode because a protective oxide layer forms on aluminium metal. This layer protects the aluminium beneath it from further corrosion.

2   Why does galvanised iron not rust even if it is scratched?

Synoptic link

Remember that steels are alloys in which the main metal is iron. Stainless steel is an alloy that does not rust (see Topic C15.2).

Key words: rusting, galvanised, sacrificial protection

Study tip

You should be able to describe experiments and interpret results to show that both air and water are necessary for rusting.

15.2 Useful alloys

Key points

- Alloys are harder than pure metals because the regular layers in a pure metal are distorted by differently sized atoms in an alloy.

- Copper, gold, and aluminium are all alloyed with other metals to make them harder.

- Pure iron is too soft for it to be very useful.

- Carefully controlled quantities of carbon and other elements are added to iron to make steel alloys with different properties.

- Important examples of steels are:

  - high carbon steels, which are very hard but brittle

  - low carbon steels, which are softer and easily shaped

  - stainless steels, which are resistant to corrosion.

Synoptic links

For more information on the structure of metals and alloys, see Topics C3.9 and C3.10.

Thin sheets of low carbon steel are easily pressed into shapes

- Alloys are mixtures of metals or metals mixed with other elements. The different sized atoms in the mixture distort the regular pattern of atoms in the layers of the metal structure and make it more difficult for them to slide over each other. This makes alloys harder than pure metals.

- Bronze is an alloy of copper and tin. Bronze is used to make statues and ships' propellers, because of its toughness and resistance to corrosion.

- Brass is an alloy of copper and zinc. Brass is much harder than copper and is used to make musical instruments, door fittings, and taps.

Brass (an alloy of copper and zinc) is used in door fittings and taps

1 Give two reasons why alloys can be more useful than pure metals.

- Aluminium has a low density for a metal. Lightweight but strong, aluminium alloys are used to build aircraft.

- Gold can be made harder by adding other elements. Gold is usually alloyed with silver, copper, and zinc in jewellery. The proportion of gold is often expressed in 'carats', where 24-carat gold is pure gold (100%). An 18-carat gold ring will contain 75% gold.

2 What percentage of gold does a 12-carat gold ring contain?

- Pure iron is too soft for many uses. Alloys of iron are called steels. By carefully controlling the amounts of carbon and other elements, the properties of steels can be changed for different uses.

- The simplest steels are the **carbon steels**. High-carbon steel, which is used in cutting tools, is very strong but brittle. Low-carbon steel is softer and more easily shaped. It can be used to make car bodies.

- Chromium–nickel steels are known as **stainless steels**. They combine hardness and strength with resistance to corrosion. These properties make them ideal for use in cutlery.

Key words: carbon steel, stainless steel

15.3 The properties of polymers

Key words: polymer, thermosoftening polymer, thermosetting polymer

- The properties of a **polymer** depend on the monomers used to make it, and the conditions used to carry out the reaction.

- There are two types of poly(ethene). Both are made from ethene monomers, but they are produced using different catalysts and reaction conditions.

- High density (HD) poly(ethene) is made from ethene using a catalyst at 50 °C and a slightly raised pressure. Ethene forms low density (LD) poly(ethene) when using very high pressures and a trace of oxygen.

- HD poly(ethene) has a higher softening temperature and is stronger than LD poly(ethene).

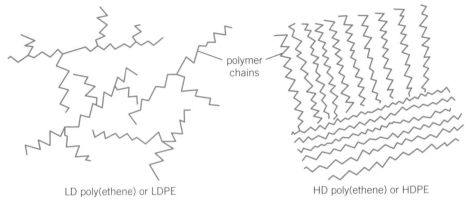

LD poly(ethene) or LDPE   HD poly(ethene) or HDPE

The branched chains of LD poly(ethene) cannot pack as tightly together as the straighter chains in HD poly(ethene), giving the polymers different properties

1   Why do LD and HD poly(ethene) have different properties?

- Poly(ethene) is an example of a **thermosoftening polymer**. It is made up of individual polymer chains that are tangled together. In thermosoftening polymers the forces between the polymer chains are weak. When you heat the polymer, these weak intermolecular forces are broken and the polymer becomes soft. When the polymer cools down, the intermolecular forces bring the polymer molecules back together so the polymer hardens again. This means it can be heated to mould it into shape and it can be remoulded by heating it again.

Extensive cross linking by covalent bonds between polymer chains makes a thermosetting plastic that is heat-resistant and rigid

- **Thermosetting polymers** do not melt or soften when heated. These polymers set hard when they are first moulded because strong covalent bonds form cross-links between their polymer chains. The strong bonds hold the polymer chains in position.

2   What is the main difference in the structures of thermosoftening and thermosetting polymers?

15.4 Glass, ceramics, and composites

Key points

- Soda-lime glass is made by heating a mixture of sand, limestone, and sodium carbonate. Borosilicate glass is made from sand and boron trioxide, and melts at a higher temperature than soda-lime glass.

- Clay ceramics include pottery and bricks. They are made by shaping wet clay then heating in a furnace.

- Composites are usually made of two materials, with one material acting as a binder for the other material, improving a desirable property that neither of the original materials could offer alone.

Key words: ceramic, composite

- Soda-lime glass is the most common form of glass. The raw materials are sand, limestone, and sodium carbonate. They are heated together to produce soda-lime glass.

- Borosilicate glass is made from sand and boron trioxide. Borosilicate glass is used for ovenware as it melts at higher temperatures than soda-lime glass.

1   What are the raw materials used to make borosilicate glass?

- Examples of **ceramic** objects made from clay are bricks, pottery, crockery, sinks, and toilets. Clay ceramics are hard, but brittle materials. They are electrical insulators and resistant to chemical attack. Ceramics are made by moulding wet clay into shapes and then heating them in a furnace.

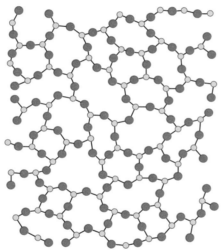

This diagram represents the disorderly structure of a glass

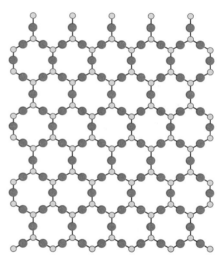

This diagram represents the orderly giant structure of a ceramic material, which has crystalline regions

- Most **composites** are made of two materials, making a product with improved properties for a particular use. They contain a matrix (or binder) of one material surrounding and binding together fibres or fragments of the other material – a process called reinforcement.

- The following materials are common composites.

  - Glass and ceramics are both brittle. When combined and heated together, they form a composite glass-ceramic, which is hard and also very tough.

  - Fibreglass is a composite of glass with polymers as the binding material. It forms a tough, flexible, waterproof material with a low density.

  - Concrete is a very hard, strong composite that is very strong in compression. It can be made more resistant to bending forces by setting it around a matrix of steel rods, forming 'reinforced concrete'.

2   Give one advantage of using reinforced concrete rather than concrete.

15.5 Making ammonia – the Haber process

Key points

- Ammonia is an important chemical for making other products, including fertilisers.
- Ammonia is made from nitrogen and hydrogen in the Haber process.
- Industrial chemists carry out the Haber process under conditions of about 450 °C and 200 atmospheres pressure, using an iron catalyst.
- Any unreacted nitrogen and hydrogen are recycled back into the reaction vessel in the Haber process.

Synoptic link

For more information about reversible reactions, see Topics C8.6 and C8.7.

Study tip

Draw a flow chart to show how ammonia is made in the Haber process.

On the up

You should be able to explain how costs are kept to a minimum in the Haber process.

- The Haber process is used to manufacture ammonia, which can be used to make fertilisers and other chemicals.
- The raw materials are nitrogen from the air and hydrogen, which is obtained from natural gas.
- The purified gases are passed over an iron catalyst at a temperature of about 450 °C and a pressure of about 200 atmospheres.
- The reaction is reversible: $N_2(g) + 3H_2(g) \rightleftharpoons 2NH_3(g)$

1   Write a word equation for the manufacture of ammonia.

- Some of the ammonia that is produced breaks down into nitrogen and hydrogen and the yield of ammonia is only about 15%.
- The gases that come out of the reactor are cooled so the ammonia condenses. The liquid ammonia is separated from the unreacted gases. The unreacted gases are recycled so they are not wasted.

2   What is done in the Haber process to conserve raw materials?

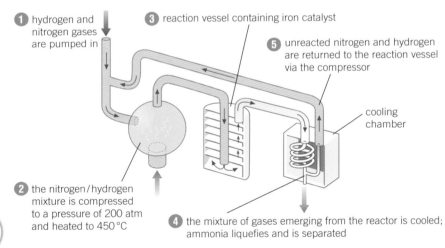

① hydrogen and nitrogen gases are pumped in

③ reaction vessel containing iron catalyst

⑤ unreacted nitrogen and hydrogen are returned to the reaction vessel via the compressor

cooling chamber

② the nitrogen / hydrogen mixture is compressed to a pressure of 200 atm and heated to 450 °C

④ the mixture of gases emerging from the reactor is cooled; ammonia liquefies and is separated

The Haber process

C15

15.6 The economics of the Haber process

Optimum pressure for the Haber process

- In the Haber process, nitrogen and hydrogen react to make ammonia in a reversible reaction:

$$N_2(g) + 3H_2(g) \rightleftharpoons 2NH_3(g)$$

- The reactants have more molecules of gas than the products, so the volume of the reactants is greater than the volume of the products. An increase in pressure will tend to shift the position of equilibrium to the right, producing more ammonia in order to reduce the pressure. This is an application of Le Châtelier's Principle.

- So the higher the pressure the greater the yield of ammonia. However, higher pressures need more energy to compress the gas and also stronger reaction vessels and pipes, which increase costs.

- A pressure of about 200 atmospheres is often used as a compromise. This pressure gives a lower yield than it would with even higher pressures, but it reduces costs and helps produce a reasonable rate of reaction between the gases.

Optimum temperature for the Haber process

- The forward reaction is exothermic and so the lower the temperature the greater the yield of ammonia. However, the reaction rate decreases as the temperature is lowered so it would take a longer time to produce any ammonia.

- Therefore, a compromise temperature of about 450 °C is usually used to give a reasonable yield in a short time.

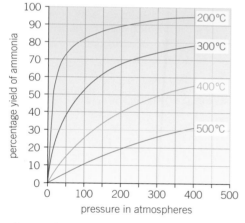

The conditions for the Haber process are a compromise between getting a reasonable yield of ammonia and getting the reaction to take place at a fast enough rate

1 At a temperature of 100 °C and 200 atmospheres pressure the yield of ammonia is 98%. Why is the Haber process not done at this temperature?

The effect of a catalyst

- An iron catalyst is also used in the Haber process to speed up the reaction. The catalyst speeds up the rate of both the forward and reverse reactions by the same amount. Therefore it does not affect the actual yield of ammonia, but it does cause ammonia to be produced more quickly, which is an important economic consideration in industry.

15.7 Making fertilisers in the lab

Key points

- Ammonia is used to make nitric acid.
- The nitric acid made can then be reacted with more ammonia to make ammonium nitrate fertiliser.
- Ammonia can also be neutralised by sulfuric acid to make ammonium sulfate fertiliser, and with phosphoric acid to make ammonium phosphate fertiliser.

- Most of the ammonia made in the Haber process is changed into compounds of ammonia. These compounds are often used as fertilisers.

- Some of the ammonia made is also converted in another process into nitric acid. Nitric acid is reacted with ammonia solution to make ammonium nitrate fertiliser:

ammonia + nitric acid → ammonium nitrate

$$NH_3(aq) + HNO_3(aq) \rightarrow NH_4NO_3(aq)$$

- You can make other fertiliser salts by reacting ammonia solution (an alkali) with different acids.

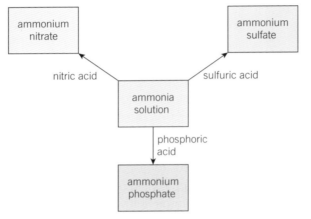

Ammonia can be made into different ammonium salts that can act as fertilisers

1 Write a word equation for the reaction between ammonia solution and sulfuric acid.

- The salts are formed in a **neutralisation** reaction. When sulfuric acid neutralises ammonia solution, ammonium sulfate is produced. You can make ammonium sulfate in the laboratory by the titration of dilute sulfuric acid against ammonia solution.

2 What method would you use to make ammonium phosphate in the laboratory from ammonia solution and phosphoric acid?

Key word: neutralisation

dilute sulfuric acid

ammonia solution

Titrating ammonia solution against dilute sulfuric acid

Synoptic link

For more information about the practical technique of titration, see Topic C4.7.

Study tip

Ammonia solution is an alkali because it contains excess hydroxide ions, OH⁻(aq), in the solution.

Using our resources

15.8 Making fertilisers in industry

Key points

- Fertilisers are used to supply nitrogen, phosphorus, and potassium to plants. These can all be added to the land at the same time in mixtures of compounds called NPK fertilisers.

- The nitrogen is provided by ammonia, made in the Haber process, which is reacted with acids to make fertilisers, such as ammonium nitrate and ammonium sulfate.

- The source of phosphorus is phosphate rock, which is mined and then treated with acids to form fertilisers, such as ammonium phosphate and calcium phosphate.

- The potassium comes from potassium salts mined from the ground for use as fertilisers, such as potassium chloride and potassium sulfate.

- Crops need the nutrients nitrogen (N), phosphorus (P), and potassium (K), for healthy growth. NPK fertilisers contain formulations of compounds to provide all three elements in appropriate proportions.

1 Name the elements found in an NPK fertiliser.

- The source of nitrogen in fertilisers is ammonia, produced in the Haber process. Ammonia is reacted with acids to make fertilisers, such as ammonium nitrate.

- The source of phosphorus is deposits of phosphate-containing rock, which are mined from the ground. The phosphate rock is insoluble in water, so it is treated with acids to make soluble fertiliser salts.

- Phosphate rock is treated with nitric acid to produce phosphoric acid and calcium nitrate. Phosphoric acid is then neutralised with ammonia to produce ammonium phosphate.

- Phosphate rock is treated with sulfuric acid to produce single superphosphate, a mixture of calcium phosphate and calcium sulfate.

- Phosphate rock is treated with phosphoric acid to produce triple superphosphate, which is calcium phosphate.

- The potassium salts, potassium chloride and potassium sulfate, are mined. They are soluble in water, so they can be separated from impurities and used directly.

2 Give one advantage of using a formulation to make a fertiliser.

On the up

You should be able to compare and contrast the industrial and laboratory production of fertilisers.

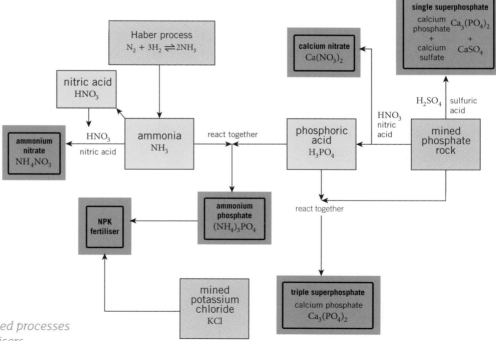

A summary of the integrated processes used to manufacture fertilisers

1. Which three substances are needed for rusting to occur? [3 marks]

2. Why are aluminium alloys used to build aircraft? [2 marks]

3. a Name the three raw materials used to make soda-lime glass. [3 marks]
   b Which of the materials is also used to make borosilicate glass? [1 mark]

4. a Name and give the formula of the fertilisers made when ammonia reacts with:
      i nitric acid   ii sulfuric acid   iii phosphoric acid. [6 marks]
   b What type of reactions are these? [1 mark]

5. Ammonia is made by the Haber process. The equation for the reaction is: $N_2(g) + 3H_2(g) \rightleftharpoons 2NH_3(g)$
   a What are the raw materials used for the process? [2 marks]
   b What conditions are used in the Haber process? [3 marks]
   c How is ammonia separated from the unreacted nitrogen and hydrogen? [3 marks]

6. a What do the letters N, P, and K stand for in NPK fertilisers? [3 marks]
   b Why are NPK fertilisers described as formulations? [4 marks]

7. a Which monomer is used to make LD poly(ethene) and HD poly(ethene)? [1 mark]
   b Why do LD poly(ethene) and HD poly(ethene) have different properties? [2 marks]

8. Write the word equation and balanced symbol equation for the reaction between ammonia solution and dilute nitric acid. [2 marks]

9. What is sacrificial protection? [3 marks]

10. Why are thermosetting polymers often used to make handles for cooking pans? [3 marks]

11. Ⓗ The forward reaction shown in the chemical equation for the Haber process in question **5** is exothermic. Explain why the Haber process is carried out at a temperature of about 450 °C. [4 marks]

Chapter checklist

Tick when you have:

reviewed it after your lesson ☑ ☐ ☐

revised once – some questions right ☑ ☑ ☐

revised twice – all questions right ☑ ☑ ☑

Move on to another topic when you have all three ticks

15.1 Rusting	☐ ☐ ☐
15.2 Useful alloys	☐ ☐ ☐
15.3 The properties of polymers	☐ ☐ ☐
15.4 Glass, ceramics, and composites	☐ ☐ ☐
15.5 Making ammonia – the Haber process	☐ ☐ ☐
15.6 The economics of the Haber process	☐ ☐ ☐
15.7 Making fertilisers in the lab	☐ ☐ ☐
15.8 Making fertilisers in industry	☐ ☐ ☐

01 The Haber process is used to produce ammonia.

$$N_2(g) + 3H_2(g) \rightleftharpoons 2NH_3(g)$$

01.1 What does the symbol $\rightleftharpoons$ tell you about the reaction? [1 mark]

01.2 What is the source of the nitrogen used in the process? [1 mark]

01.3 Why is iron powder used in the reactor? [1 mark]

01.4 What happens to the unreacted nitrogen and hydrogen? [1 mark]

01.5 How are the conditions changed in the separator so the ammonia liquefies? [2 marks]

02 Sodium ions can be detected by both flame tests and flame emission spectroscopy.

02.1 What is the colour produced by sodium ions in a flame? [1 mark]

02.2 Name another element in Group 1 that gives a colour in the flame and state the colour. [2 marks]

02.3 How can flame emission spectroscopy be used to find the concentration of an element in solution? [1 mark]

02.4 Why can flame emission spectroscopy be used to identify elements? [1 mark]

02.5 Suggest **three** reasons why flame emission spectroscopy is more useful than flame tests in industry. [3 marks]

03 Limestone is mainly calcium carbonate, $CaCO_3$. When limestone is heated strongly the calcium carbonate breaks down.

03.1 Write a word equation for this reaction. [1 mark]

03.2 What type of reaction is this? [1 mark]

03.3 When water is added to the solid product, calcium hydroxide, $Ca(OH)_2$, is produced. Adding more water and filtering produces a solution of calcium hydroxide. Universal indicator solution was added to calcium hydroxide solution. The indicator turned blue. What does this tell you about calcium hydroxide? [1 mark]

03.4 When carbon dioxide is bubbled into calcium hydroxide solution, the solution turns cloudy. Explain why. [2 marks]

04 Copper can be extracted from copper-rich ores by smelting. This is done by heating the ore in a furnace to about 1100 °C and blowing air through it. An equation for a reaction in the furnace is:

copper sulfide + oxygen → copper + sulfur dioxide

04.1 Copper ore and air are needed for this process. What other resource is needed? [1 mark]

04.2 Why should sulfur dioxide not be allowed to escape into the air? [1 mark]

04.3 The copper that is produced is impure. Name the method used to purify the copper. [1 mark]

05 **H** Copper can be extracted from low-grade ores by bioleaching. A solution containing water, bacteria, and sulfuric acid is added to the top of a heap of ore. The leachate solution that is collected from the bottom of the heap contains copper sulfate. Copper is extracted from the solution using scrap iron. The solution can be re-used.

05.1 What is the purpose of the bacteria in this process? [1 mark]

05.2 Write a word equation for the reaction between iron and copper sulfate. [1 mark]

05.3 What is the name of this type of reaction? [1 mark]

05.4 The solution from which the copper has been extracted should not be allowed to escape without further treatment. Explain why. [2 marks]

Study tip

When a question uses the word 'suggest' as in question **02.5**, there may be lots of possible answers. You only need to give the number that the question asks for, but make sure that they are relevant.

Study tip

Many questions will give you some information or data. It is important that you read this carefully and that your answer relates to this information and not some other context that you have studied.

06 **Table 1** shows the percentages of the four most abundant gases in dry air.

Table 1

Name of gas	Percentage by volume in dry air
nitrogen	78.08
oxygen	20.95
argon	0.93
carbon dioxide	0.03

06.1 Which of these gases is believed to have been the most abundant in the Earth's early atmosphere? [1 mark]

06.2 Name **one** other gas that was probably in the Earth's early atmosphere. [1 mark]

06.3 What produced the oxygen that is in the air? [2 marks]

06.4 Why are there many theories about how life began on Earth? [1 mark]

06.5 **Table 2** shows the boiling points of the three most abundant elements in air. [2 marks]

Table 2

Name of element	Boiling point in °C
argon	−186
nitrogen	−196
oxygen	−183

To separate these elements, air is cooled to −200 °C, so that the gases become liquids. Suggest how the liquid mixture of gases can be separated.

06.6 Explain how the process chosen in **06.5** will separate the elements shown. [5 marks]

07 Ammonium phosphate $(NH_4)_3PO_4$ can be used as a fertiliser.

07.1 Calculate the relative formula mass of ammonium phosphate.
(A_r of N = 14, A_r of H = 1, A_r of O = 16, A_r of P = 31) [2 marks]

07.2 Ammonium phosphate contains atoms of the isotope $^{31}_{15}P$. Another isotope of phosphorus is $^{32}_{15}P$.
Explain what is meant by the term isotope using these two types of phosphorus atoms as examples. [3 marks]

07.3 Describe how you would make ammonium phosphate crystals in the laboratory. [6 marks]

08 Ⓗ The Haber process is used to produce ammonia.
$$N_2(g) + 3H_2(g) \rightleftharpoons 2NH_3(g)$$

08.1 Explain why a high pressure is used for the reaction. [2 marks]

08.2 Explain why the yield of ammonia decreases with an increase in temperature. [2 marks]

08.3 A relatively high temperature of 450 °C is used in the process. Explain why. [4 marks]

09 For each of the following pairs of substances, suggest **one** chemical test that you could do to tell them apart. Give the results of the test for both substances.

09.1 sodium carbonate and sodium nitrate [3 marks]

09.2 potassium chloride and potassium iodide [3 marks]

09.3 calcium chloride and magnesium chloride [3 marks]

09.4 iron(II) sulfate and iron(III) sulfate [3 marks]

09.5 ethanol and ethanoic acid [3 marks]

Study tip

Some longer questions, such as question **07.3** may have a level of response mark scheme. In these mark schemes, there is not a mark for each scientific response that you make. It is the overall quality of the answer that you write that decides the mark.

Study tip

When asked for a chemical test, as in question **09**, you should always describe how to do the test and give the result of the test.

Calcium carbonate reacts with dilute hydrochloric acid. The equation for the reaction is

$$CaCO_3(s) + 2\,HCl(aq) \rightarrow CaCl_2(aq) + H_2O(l) + CO_2(g)$$

A student investigated how changing the concentration of dilute hydrochloric acid altered how quickly carbon dioxide gas was made.

The gas was collected over water and its volume measured every 20 seconds.

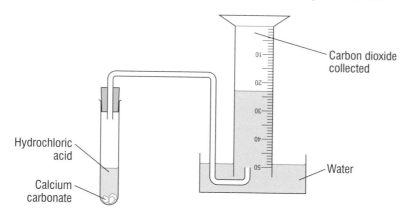

Introduction

Below is an example of an exam-style question focusing on practical skills. The student answers are marked and include comments to help you answer this type of exam question as effectively as possible.

1 a   The diagram shows the apparatus during one experiment.

What volume of carbon dioxide gas has been collected?

23 cm³

(1)

> The volume is correct and units have been given. Level has been read to the bottom of the meniscus.

The graph shows the results the student obtained.

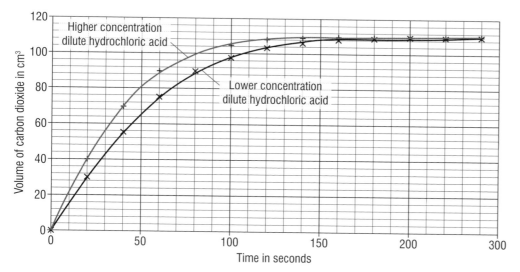

> There are two marks available for the answer, so there will normally be two things that need to be said to gain full marks. This candidate has spotted that while the final volume of gas collected is the same (the graphs level off at the same height) the line for the more concentrated acid is steeper – this means that at any time before the reaction is complete, more gas has been collected when the more concentrated acid is used.

b   Describe the effect of changing the concentration of the dilute hydrochloric acid on the volume of carbon dioxide collected during the experiment.

When the concentration of the acid is higher the gas is collected more quickly but the same final volume of gas is collected.

(2)

c   Carbon dioxide gas is slightly soluble in water. What effect will this have on the volume of gas collected?

It will reduce the volume of gas collected.

(1)

> Due to the fact that carbon dioxide is slightly soluble in water, a small amount of it will dissolve in the water in the water bath. This will reduce the volume of gas that can be collected in the measuring cylinder.

A variable which must be controlled has been identified (other variables that must be controlled include the temperature of the acid). This scores one mark. The remaining two marks are awarded for explaining why changing the selected variable would change the results.

d The student carried out her investigation so that it was a fair test. Select one variable that must be kept constant for the experiment to be a fair test and explain why it must be kept constant.

The surface area of the calcium carbonate must be kept constant. If the surface area was higher, then there would be more collisions of hydrogen ions with calcium carbonate per second. This would make the reaction faster.

(3)

Another student investigated the mass of calcium carbonate used on the final volume of carbon dioxide gas collected. The method she followed was:

1. Place 25.0 cm³ of dilute hydrochloric acid in a boiling tube.
2. Add a known mass of calcium carbonate to the dilute acid in the boiling tube.
3. Connect the boiling tube to a gas syringe as shown in the diagram

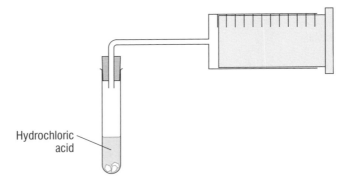

Hydrochloric acid

The graph shows the results she obtained.

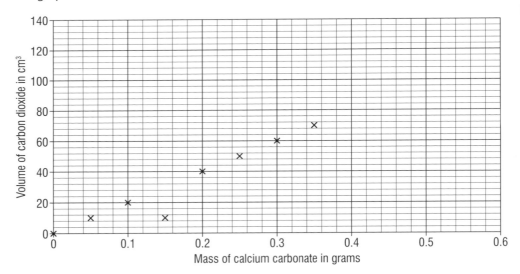

A volumetric pipette is the best apparatus to measure a fixed volume such as 25.0 cm³; a burette could also be used.

e Name an item of laboratory apparatus most suitable for measuring 25.0 cm³ of hydrochloric acid.

A volumetric pipette could be used.

(1)

f   Draw a line of best fit through the points on the graph.

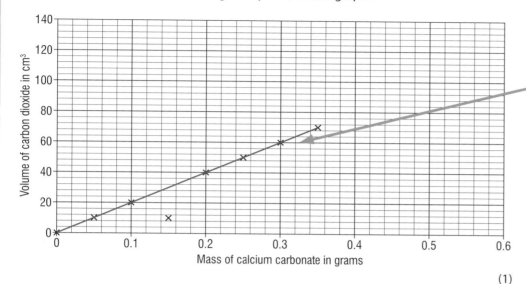

(1)

The line drawn is straight (it has been done with the aid of a rule) and it ignores the anomalous point.

g   Describe the relationship between the mass of calcium carbonate used and the volume of carbon dioxide gas collected.

As the mass of calcium carbonate is doubled, the volume of gas collected also doubles.

(2)

An alternative way to say this is "they are directly proportional". A straight line through the origin shows a directly proportional relationship. If the line is straight but does not pass through the origin then we can say it is a linear relationship (but it is not directly proportional).

h   There is an anomalous point on the graph. Circle the anomalous point and explain what could have happened to cause the anomalous point.

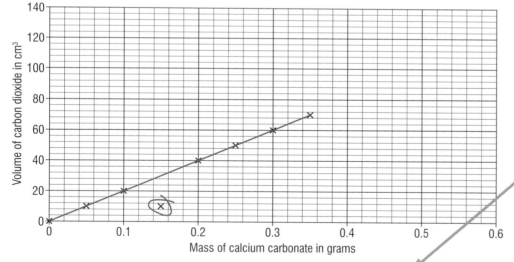

The bung may not have been replaced correctly. This would have allowed some of the gas to escape, and so less was collected.

(3)

The point that is some distance from the line has been correctly identified. The explanation gives a correct reason why the volume of gas collected could have been too small and explains why the reason given would result in too small a volume. Other possible explanations include the mass of calcium carbonate being too small so less carbon dioxide would be made.

i   What should the student have done to check the precision of her results and improve the accuracy of the data she has plotted?

The experiment should have been repeated at least three times and then a mean value calculated for the volume of gas collected for each mass of calcium carbonate.

(2)

Repeating the experiment does not necessarily make the results more accurate, but it does let us check to see if the results are precise. Calculating a mean using the repeated results will reduce any random errors and so the mean values should be more accurate.

j Use the graph to predict the volume of gas that would be collected if 0.55 g of calcium carbonate were used in the experiment. Show all of your working.

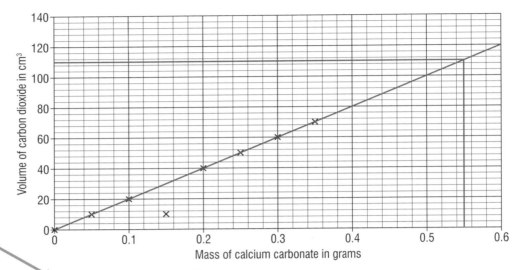

This scores both marks because the candidate has extrapolated the graph line and then read off the correct value from the graph. The value has been read from the graph accurately.

The volume of gas would be 110 cm³.

(2)

The student then repeated the experiment using 0.55 g of calcium carbonate. She obtained a much smaller volume of gas than she expected. Her teacher suggested that this was because she had not used enough acid to react with all the calcium carbonate

k Explain how she could check whether her teacher's suggestion was correct.

She could repeat the experiment using a larger volume of acid.

If the teacher is correct then the volume of gas will increase.

(2)

This scores both marks as the candidate has said what should be done and the result that should be obtained if the teacher is correct.

The student calculated the volume of gas, at 25 °C, she should have made in some of the runs. The table shows the volumes she collected and how much should have been made.

All of the gas volumes are too small. This means there is a problem with the procedure being used.

Mass calcium carbonate/g	Volume carbon dioxide collected/cm³	Theoretical volume of carbon dioxide/cm³
0.10	20	24
0.20	40	48
0.30	60	72

l What type of error do her results show?

It is a systematic error.

(1)

This answer scores both marks as it gives a problem that would cause a systematic error and explains how that error makes the gas volume too small. Another correct answer would be if the gas was being collected at a temperature below 25°C as at lower temperatures the gas will have a smaller volume.

m Explain how an error of this type could have occurred in each run.

As soon as the calcium carbonate has been added to the acid carbon dioxide gas is made. Some of this gas escapes before the bung is placed in the boiling tube connected to the gas syringe.

(2)

Practical questions

01 A student used the apparatus shown in **Figure 1** to investigate the temperature rise when metals were added to hydrochloric acid.

01.1 Name the instrument the student used to measure temperature. [1 mark]

01.2 The student used the same volume of acid in each test.
What type of variable is this? Choose **two** correct letters. [1 mark]
   A categoric   **B** continuous   **C** control

01.3 Name an instrument the student could use to measure the volume of acid. [1 mark]

01.4 Name the independent variable in this investigation. [1 mark]

01.5 Name the dependent variable in this investigation. [1 mark]

01.6 The teacher demonstrated the process using calcium metal.
The teacher's results were:

Starting temperature	21.3 °C
Maximum temperature	44.5 °C

Calculate the temperature rise for calcium metal. [1 mark]

01.7 **Table 1** shows the student's results.
How would you display the results? Choose the correct letter. [1 mark]
   D bar chart   **E** line graph   **F** pie chart   **G** scatter graph

01.8 Explain your choice of display method in **01.7**. [1 mark]

01.9 Place the metals in **Table 1** in order of reactivity. [2 marks]

01.10 Predict the temperature rise when lithium is added to hydrochloric acid. [1 mark]

01.11 Explain why the student would not use lithium in this investigation. [2 marks]

01.12 Give **two** reasons why the student should repeat the investigation. [2 marks]

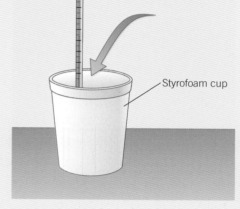

Figure 1

Table 1

Metal	Temperature rise in °C
copper	0.0
iron	8.3
magnesium	16.5
zinc	12.0

02 A student measured the maximum mass of potassium chloride that would dissolve in $100\,cm^3$ of water at different temperatures.
Table 2 shows the student's results.

Table 2

Maximum mass of potassium chloride in grams that will dissolve in 100 cm³ of water	Temperature in °C
31.2	10
34.2	20
37.2	30
40.1	40
44.6	50
45.8	60

02.1 Plot the results shown in **Table 2** on a graph.
On the x-axis, plot temperature in °C. On the y-axis, plot mass in grams. Start the y-axis at 30.0 g.
Draw a line of best fit through the points. [4 marks]

02.2 There is an anomalous point. Circle the anomalous point on the graph drawn in **02.1**. [1 mark]

02.3 Suggest the mass of potassium chloride that would dissolve in $100\,cm^3$ of water at 70 °C. Show your working on the graph drawn in **02.1**. [2 marks]

02.4 How would the student know when the maximum mass of potassium chloride had dissolved? [1 mark]

02.5 All measurements have some degree of error. The student used a thermometer with scale divisions every 2 °C, so the uncertainty of each reading of the temperature can be taken as ± 1 °C.
Calculate the percentage error in the measurement of the temperature read as 40 °C in the experiment. [1 mark]

03 Students investigated three dyes, **A**, **B**, and **C**.
One student set the apparatus up as shown in **Figure 2**.

03.1 Name **two** mistakes the student made in setting up the apparatus.
For each mistake, state the effect it would have. [4 marks]

03.2 Another student set up the investigation correctly.
The student's results are shown in **Figure 3**.
Give **three** conclusions you can make about dye **A**. [3 marks]

03.3 Calculate the R_f of dye **B**. Give your answer to 3 significant figures. [5 marks]
Use the equation below to help you.

$$R_f = \frac{\text{distance moved by dye}}{\text{distance moved from start line to solvent front}}$$

04 A student investigated the reaction of marble chips and dilute
hydrochloric acid. Marble chips consist mainly of calcium carbonate.
The equation for the reaction is:
$$CaCO_3(s) + 2HCl(aq) \rightarrow CaCl_2(aq) + H_2O(l) + CO_2(g)$$
Figure 4 shows the apparatus the student used.
The student measured the mass lost in grams from the flask and its
contents every 15 seconds. **Table 3** shows the student's results.

04.1 Plot these results on a graph. Draw a line of best fit through
the points. [4 marks]

04.2 Describe the pattern shown by the graph. [2 marks]

04.3 What causes the loss in mass? [1 mark]

04.4 Give **two** observations the student would make during the
investigation. [2 marks]

04.5 Give **two** conclusions you can make from the graph drawn in
04.1 about the rate of reaction. [2 marks]

04.6 Suggest a different method the student could use to follow
the rate of reaction. [2 marks]

04.7 The student repeated the experiment at a higher temperature.
Sketch on the graph drawn in **04.1** the results you would expect
the student to obtain. [2 marks]

05 A student investigated the volume of hydrochloric acid needed
to react with $25\,cm^3$ of $0.10\,mol/dm^3$ of sodium hydroxide. The sodium
hydroxide and an indicator were placed in a conical flask.
The student added dilute hydrochloric acid from a burette until the
indicator changed colour.

05.1 Suggest **one** variable the student would need to control to make
sure that the results are valid. [1 mark]

05.2 **Table 4** shows the measurements the student made.

Table 4

Burette reading in cm³	Test 1	Test 2	Test 3	Test 4
Final burette reading	22.90	45.20	22.25	42.25
Initial burette reading		23.00	0.00	22.25
Volume of acid added				

Figure 5 shows the volume of acid in the burette at the start of
Test 1.
Give the reading that would be entered in **Table 4**. [1 mark]

05.3 Concordant results are within $0.10\,cm^3$ of each other.
Use the student's concordant results to work out the mean
volume of hydrochloric acid added. Give your answer to the
appropriate number of decimal places. [5 marks]

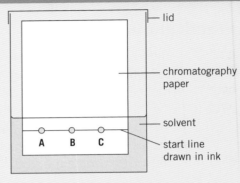

Figure 2

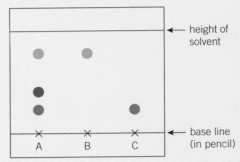

Figure 3

Figure 4

Table 3

Time in s	Mass lost in g
0	0
15	1.2
30	1.8
45	2.1
60	2.2
75	2.3
90	2.3

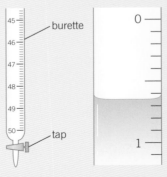

Figure 5

Summary of 'Working scientifically'

A minimum of 15% of the total marks in your exam will be based on 'Working scientifically'. Questions could ask about the methods and techniques you have practised during the Required practicals (see the Practice questions on pages 123–4). You will also be asked to apply the principles of scientific enquiry in general, as outlined below.

WS1 Development of scientific thinking

- Understand how scientific methods and theories develop over time.
- Use a variety of models to solve problems, make predictions, and to develop scientific explanations and understanding of familiar and unfamiliar facts.
- Appreciate the power and limitations of science and consider any ethical issues that may arise.
- Explain everyday and technological applications of science. This will include evaluating their personal, social, economic, and environmental implications. You may also be asked to make decisions based on the evaluation of evidence and arguments.
- Evaluate risks both in practical science and in wider contexts, including perception of risk in relation to data and consequences.
- Recognise the importance of peer review of results and of communicating results to a range of audiences.

WS2 Experimental skills and strategies

- Use scientific theories and explanations to develop hypotheses.
- Plan experiments or devise procedures to make observations, produce or characterise a substance, test hypotheses, check data, or explore phenomena.
- Apply knowledge of a range of techniques, instruments, apparatus, and materials to select those appropriate to the experiment.
- Carry out experiments, handling apparatus correctly and taking into account the accuracy of measurements, as well as any health and safety considerations.
- Recognise when to apply knowledge of sampling techniques to make sure that any samples collected are representative.
- Make and record observations and measurements using a range of apparatus and methods.
- Evaluate methods and suggest possible improvements and further investigations.

WS3 Analysis and evaluation

- Present observations and other data using appropriate methods.
- Translate data from one form to another.
- Carry out and represent mathematical and statistical analysis.
- Represent distributions of results and make estimates of uncertainty.
- Interpret observations and other data, including identifying patterns and trends, making inferences, and drawing conclusions.
- Present reasoned explanations, including relating data to hypotheses.
- Be objective; evaluating data in terms of accuracy, precision, repeatability, and reproducibility; and identifying potential sources of random and systematic error.
- Communicate the reasoning for investigations, the methods used, the findings, and reasoned conclusions through paper-based and electronic reports, as well as using other forms of presentation.

WS4 Scientific vocabulary, quantities, units, symbols, and nomenclature

- Use scientific vocabulary, terminology, and definitions.
- Recognise the importance of scientific quantities and understand how they are determined.
- Use SI units (e.g., kg, g, mg; km, m, mm; kJ, J) and IUPAC chemical names, whenever appropriate.
- Use prefixes and powers of ten for orders of magnitude (e.g., tera, giga, mega, kilo, centi, milli, micro, and nano).
- Interconvert units.
- Use an appropriate number of significant figures in calculation. Quote your answer to the same number of significant figures as the data provided in the question, to the least number of significant figures.

Essential 'Working scientifically' terms

Your knowledge of what it means to 'work scientifically' will be tested in your exams. In order to understand the nature of science and experimentation, you will need to know some technical terms that scientists use. You should be able to recognise and use the terms below:

accurate a measurement is considered accurate if it is judged to be close to the true value

anomalies/anomalous results results that do not match the pattern seen in the other data collected or that are well outside the range of other repeat readings (outliers)

categoric variable has values that are labels (described in words), for example, types of material

control variable a variable that may, in addition to the variable under investigation (the independent variable – see below), affect the outcome of an investigation and therefore has to be kept constant, or at least has to be monitored as carefully as possible

data information, either qualitative (descriptive) or quantitative (measured), that has been collected

dependent variable the variable for which the value is measured for each and every change in the variable under investigation (called the independent variable – see below)

directly proportional a relationship that, when drawn on a line graph, shows a positive linear relationship (a straight line) that crosses through the origin

fair test a test in which only the independent variable has been allowed to affect the dependent variable

gradient (of a straight line graph) change of the quantity plotted on the y-axis divided by the change of the quantity plotted on the x-axis

hazards anything that can cause harm, for example, an object, a property of a substance, or an activity

hypothesis a proposal intended to explain certain facts or observations

independent variable the variable under investigation, for which values are changed or selected by the investigator

line graph used when both variables (x and y) plotted on a graph are continuous. The line should normally be a line of best fit, and may be straight or a smooth curve

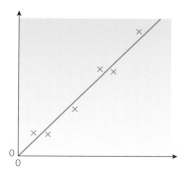

When a straight line of best fit goes through the origin (0, 0) the relationship between the variables is directly proportional

peer review evaluation of scientific research before its publication by others working in the same field

precise a precise measurement is one in which there is very little spread about the mean value. Precision depends only on the extent of random errors – it gives no indication of how close results are to the true (accurate) value

precision a measure of how precise a measurement is

prediction a forecast or statement about the way in which something will happen in the future, which is often quantitative and based on a theory or model

random error an error in measurement caused by factors that vary from one measurement to another

range the maximum and minimum values for the independent or dependent variables – this is important in ensuring that any patterns are detected and are valid

relationship the link between the variables (the independent and dependent variables, x and y) that were investigated

repeatable a measurement is repeatable if the original experimenter repeats the investigation using the same method and equipment and obtains the same results or results that show close agreement

reproducible a measurement is reproducible if the investigation is repeated by another person using different equipment and the same results are obtained

resolution this is the smallest change in the quantity being measured by a measuring instrument that gives a perceptible change in the reading

risk depends on both the likelihood of exposure to a hazard and the seriousness of any resulting harm

systematic error error that causes readings to be spread about some value other than the true value, due to results differing from the true value by a consistent amount each time a measurement is made. Sources of systematic error can include the environment, methods of observation, or instruments used. Systematic errors cannot be dealt with by simple repeats. If a systematic error is suspected, the data collection should be repeated using a different technique or a different set of equipment, and the results compared

tangent a straight line drawn to touch a point on a curve so that it has the same gradient as the curve at that point

validity the suitability of the investigative procedure to answer the question being asked

uncertainty the degree of variability that can be expected in a measurement. A reasonable estimate of the uncertainty in a mean value calculated from a set of repeat readings would be half of the range of the repeats. In an individual measurement, its uncertainty can be taken as half the smallest scale division marked on the measuring instrument or half the last figure shown on the display on a digital measuring instrument

variables physical, chemical, or biological quantities or characteristics

Glossary

Acid When dissolved in water, its solution has a pH value less than 7. Acids are proton (H^+ ion) donors.

Activation energy The minimum energy needed for a reaction to take place.

Addition polymerisation The reaction in which unsaturated monomers (containing $C{=}C$ bonds) join together to make a polymer as the only product.

Alkali metals Elements in Group 1 of the periodic table.

Alkali Its solution has a pH value more than 7.

Alkane Saturated hydrocarbon with the general formula C_nH_{2n+2}, for example, methane, ethane, and propane.

Alkene Unsaturated hydrocarbon that contains a carbon–carbon double bond. Its general formula is C_nH_{2n}, for example, ethene, C_2H_4.

Alloy A mixture of two or more elements, at least one of which is a metal.

Anhydrous Describes a substance that does not contain water.

Anode The positive electrode in electrolysis.

Aqueous solution The mixture made by adding a soluble substance to water.

Atmosphere The relatively thin layer of gases that surrounds planet Earth.

Atom The smallest part of an element that can still be recognised as that element.

Atom economy A measure of the amount of starting materials that end up as useful products.

Atomic number The number of protons (which equals the number of electrons) in an atom. It is sometimes called the proton number.

Avogadro constant The number of atoms, molecules, or ions in a mole of any substance (i.e., 6.02×10^{23} per mol).

Balanced symbol equation A symbol equation in which there are equal numbers of each type of atom on either side of the equation.

Base The oxide, hydroxide, or carbonate of a metal that will react with an acid, forming a salt as one of the products.

(If a base dissolves in water it is called an alkali). Bases are proton (H^+ ion) acceptors.

Bioleaching The process of extracting metals from ores using microorganisms.

Biofuel Fuel made from animal or plant products.

Blast furnace A huge reaction vessel used in industry to extract iron from its ore.

Bond energy The energy required to break a specific chemical bond.

Burette A long glass tube with a tap at one end and markings to show volumes of liquid; used to add precisely known volumes of liquids to a solution in a conical flask below it.

Carbon capture and storage A method of pumping carbon dioxide underground and storing it in disused oil and gas fields, thereby preventing it from escaping into the atmosphere.

Carbon footprint The total amount of carbon dioxide and other greenhouse gases emitted over the full life cycle of a product, service, or event.

Carbon steels Alloys of iron containing controlled, small amounts of carbon.

Catalyst A substance that speeds up a chemical reaction by providing a different pathway for the reaction that has a lower activation energy. The catalyst is chemically unchanged at the end of the reaction.

Cathode The negative electrode in electrolysis.

Ceramic Material made by heating clay, or other compounds, to high temperatures (called firing) to make hard, but often brittle, materials, which make excellent electrical insulators.

Chromatography The process whereby small amounts of dissolved substances are separated by running a solvent along a material, such as absorbent paper.

Climate change The change in global weather patterns that could be caused by excess levels of greenhouse gases in the atmosphere.

Closed system A system in which no matter enters or leaves.

Collision theory An explanation of chemical reactions in terms of reacting particles colliding with sufficient energy for a reaction to take place.

Composites Materials made of two or more different materials, containing a matrix or binder surrounding and binding together fibres or fragments of another material which acts as the reinforcement.

Compound A substance made when two or more elements are chemically bonded together.

Concentration The amount of substance (usually in g or mol) dissolved in 1 cubic decimetre (dm^3) of solution.

Condensation polymerisation The reaction in which two monomers with two different functional groups join together to form a polymer plus a small molecule (such as H_2O or HCl) as the other product.

Covalent bond The bond between two atoms that share one or more pairs of electrons.

Covalent bonding The attraction between two atoms that share one or more pairs of electrons.

Cracking The reaction used in the oil industry to break down large hydrocarbons into smaller, more useful ones.

Delocalised electron A bonding electron that is no longer associated with any one particular atom.

Displacement reaction A reaction in which a more reactive element takes the place of a less reactive element in one of its compounds or in solution.

Distillation Separation of a liquid from a mixture by evaporation followed by condensation.

Dot and cross diagram A drawing to show only the arrangement of the outer shell electrons of the atoms or ions in a substance.

DNA (deoxyribonucleic acid) A large organic molecule that encodes genetic instructions for the development and functioning of living organisms and viruses.

Glossary

Double bond A covalent bond made by the sharing of two pairs of electrons.

Electrolysis The breakdown of a substance containing ions by electricity.

Electrolyte A liquid, containing free-moving ions, which is broken down by electricity in the process of electrolysis.

Electron A tiny particle with a negative charge. Electrons orbit the nucleus of atoms or ions in shells.

Electronic structure A set of numbers to show the arrangement of electrons in their shells (or energy levels).

Element A substance made up of only one type of atom. An element cannot be broken down chemically into any simpler substance.

End point The point in a titration where the reaction is complete and titration should stop.

Endothermic A reaction that takes in energy from the surroundings.

Equilibrium The point in a reversible reaction at which the forward and backward rates of reaction are the same. Therefore, the amounts of substances present in the reacting mixture remain constant.

Exothermic A reaction that transfers energy to the surroundings.

Fermentation The reaction in which the enzymes in yeast turn glucose into ethanol and carbon dioxide.

Filtration The technique used to separate substances that are insoluble in a particular solvent from those that are soluble.

Flame emission spectroscopy A method of instrumental analysis in which the light given off when a sample is placed in a flame produces characteristic line spectra to identify and measure the concentration of metal ions in the sample.

Flammable Easily ignited and capable of burning rapidly.

Formulation A mixture that has been designed as a useful product.

Fractions Hydrocarbons with similar boiling points separated from crude oil.

Fractional distillation A way to separate liquids from a mixture of liquids by boiling off the substances at different temperatures, then condensing and collecting the liquids.

Fuel cells Sources of electricity that are supplied by an external source of fuel.

Fullerenes Forms of the element carbon that can exist as large cage-like structures, based on hexagonal rings of carbon atoms.

Functional group An atom or group of atoms that give organic compounds their characteristic reactions.

Galvanised Iron or steel objects that have been protected from rusting by a thin layer of zinc metal at their surface.

Giant covalent structure A huge 3D network of covalently bonded atoms.

Giant lattice / structure A huge 3D network of atoms or ions.

Gradient A measure of the slope of a straight line on a graph.

Group All the elements in the columns (labelled 1 to 7 and 0) in the periodic table.

Half equation An equation that describes reduction (gain of electrons) or oxidation (loss of electrons).

Halogens The elements found in Group 7 of the periodic table.

Homologous series A group of related organic compounds that have the same functional group.

Hydrated Describes a substance that contains water in its crystals.

Hydrocarbon A compound containing only hydrogen and carbon.

Incomplete combustion When a fuel burns in insufficient oxygen, producing carbon monoxide as a toxic product.

Inert Unreactive.

Intermolecular forces The attraction between the individual molecules in a covalently bonded substance.

Ion A charged particle produced by the loss or gain of electrons.

Ionic bond The electrostatic force of attraction between positively and negatively charged ions.

Ionic equation An equation that shows only those ions or atoms that change in a chemical reaction.

Isotopes Atoms that have the same number of protons but different number of neutrons (i.e., they have the same atomic number but different mass numbers).

Law of conservation of mass The total mass of the products formed in a reaction is equal to the total mass of the reactants.

Le Châtelier's Principle When a change in conditions is introduced to a system at equilibrium, the position of equilibrium shifts so as to cancel out the change.

Life cycle assessments (LCA) Carried out to assess the environmental impact of products, processes, or services at different stages in their life cycle.

Limiting reactant The reactant that is used up first in a reaction.

Mass number The number of protons plus neutrons in the nucleus of an atom.

Mixture When some elements or compounds are mixed together and intermingle but do not react together (i.e. no new substance is made). A mixture is not a pure substance.

Mole The amount of substance in the relative atomic or formula mass of a substance in grams.

Monomers Small reactive molecules that react together in repeating sequences to form a very large molecule (a polymer).

Nanoscience The study of very tiny particles or structures between 1 and 100 nanometres in size, where 1 nanometre = 10^{-9} metres.

Neutral A solution with a pH value of 7 which is neither acidic nor alkaline. Alternatively, something that carries no overall electrical charge.

Neutralisation The chemical reaction of an acid with a base in which a salt and water are formed. If the base is a carbonate or hydrogen carbonate, carbon dioxide is also produced in the reaction.

Neutron A dense particle found in the nucleus of an atom. It is electrically neutral, carrying no charge.

Nitrogen oxides Gaseous pollutants given off from motor vehicles, a cause of acid rain.

Noble gases The very unreactive gases found in Group 0 of the periodic table. Their atoms have very stable electronic structures.

Non-renewable Something that cannot be replaced once it is used up.

Nucleotides Monomers that undergo condensation polymerisation to form DNA.

Nucleus (of an atom) The very small and dense central part of an atom that contains protons and neutrons.

Ore Rock that contains enough metal to make it economically worthwhile to extract the metal.

Oxidation A reaction in which oxygen is added to a substance or electrons are lost.

Particle theory Explains the properties of materials by describing the movement of and distance between their particles.

Particulate Small, solid particle given off from motor vehicles as a result of incomplete combustion of their fuel.

Percentage yield The actual mass of product collected in a reaction divided by the maximum mass that could have been formed in theory, multiplied by 100.

Periodic table An arrangement of elements in the order of their atomic numbers, forming groups and periods.

pH scale A numeric scale that shows how strongly acidic or alkaline a solution is.

Pipette A glass tube used to measure accurate volumes of liquids.

Polymer A substance made from very large molecules made up of many repeating units.

Product A substance made as a result of a chemical reaction.

Proteins Natural polymers made from different amino acid monomers.

Proton A tiny positive particle found inside the nucleus of an atom. Reactant A substance we start with before a chemical reaction takes place.

Reaction profile The relative difference in the energy of reactants and products.

Reactivity series A list of elements in order of their reactivity.

Recycling The process in which waste materials are processed to be used again.

Reduction A reaction in which oxygen is removed or electrons are gained.

Relative atomic mass (A_r) The average mass of the atoms of an element compared with carbon-12 (which is given a mass of exactly 12). The average mass must take into account the proportions of the naturally occurring isotopes of the element.

Relative formula mass (M_r) The total of the relative atomic masses, added up in the ratio shown in the chemical formula, of a substance.

Reversible reaction A reaction in which the products can re-form the reactants.

R_f (retention factor) A measurement from chromatography: it is the distance a spot of substance has been carried above the baseline divided by the distance of the solvent front.

Rusting The corrosion of iron.

Sacrificial protection An effective way to prevent rusting whereby a metal more reactive than iron (such as zinc or magnesium) is attached to or coated on an object.

Salt A compound formed when some or all of the hydrogen in an acid is replaced by a metal.

Saturated hydrocarbon A hydrocarbon with only single bonds between its carbon atoms. This means that it contains as many hydrogen atoms as possible in each molecule.

Shell An area in an atom, around its nucleus, where electrons are found.

Stainless steel A chromium-nickel alloy of steel which does not rust.

State symbols The abbreviations used in balanced symbol equations to show if reactants and products are solid (s), liquid (l), gas (g) or dissolved in water (aq).

Strong acids Acids that completely ionise in solution and have a high concentration of $H^+(aq)$ ions in solution.

Thermal decomposition The breakdown of a compound by heating it.

Thermosetting polymer A polymer that can form extensive cross-linking between chains, resulting in rigid materials which are heat-resistant.

Thermosoftening polymer A polymer that forms plastics that can be softened by heating, then remoulded into different shapes as they cool down and set.

Titration A method for measuring the volumes of two solutions that react together.

Transition elements Elements from the central block of the periodic table.

Universal indicator A mixture of indicators that can change through a range of colours to show how strongly acidic or alkaline liquids and solutions are.

Unsaturated hydrocarbon A hydrocarbon whose molecules contain at least one carbon–carbon double bond.

Viscosity The resistance of a liquid to flowing or pouring; a liquid's 'thickness'.

Weak acids Acids that do not ionise completely in aqueous solutions.

Word equation A way of describing what happens in a chemical reaction by showing the names of all reactants and the products they form.

Yield The amount of product formed in a chemical reaction.

Answers

C1.1 Atoms
1. elements
2. a compound

C1.2 Chemical equations
1. hydrogen + oxygen → water
2. $H_2 + Br_2 \rightarrow 2HBr$
3. (g)
4. law of conservation of mass

C1.3 Separating mixtures
1. filtration, crystallisation, distillation
2. filtration

C1.4 Fractional distillation and paper chromatography
1. liquid with the lowest boiling point
2. paper chromatography

C1.5 History of the atom
1. J.J. Thomson

C1.6 Structure of the atom
1. 13 protons, 13 electrons, 14 neutrons

C1.7 Ions, atoms, and isotopes
1. by losing electrons
2. 12
3. have a different number of neutrons

C1.8 Electronic structures
1. diagram: three concentric circles with dot or Al at centre, innermost circle with two electrons (dots or crosses), next with eight electrons, outer circle with three electrons
2. 2,8,4
3. two

C1 Summary questions
1. a calcium; [1] hydrogen; [1] neon; [1] oxygen [1]
   b compounds [1]
2. a $H_2 + Cl_2 \rightarrow 2HCl$ [1]
   b hydrogen + chlorine → hydrogen chloride [1]
3. fractional distillation [1]
4. in order of atomic (proton) number [1]
5. a proton; [1] neutron; [1] electron [1]
   b proton: relative mass 1; [1] relative charge +1 [1]
   neutron: relative mass 1; [1] relative charge 0 [1]
   electron: relative mass very small; [1] relative charge −1 [1]
6. heat the sodium chloride solution [1] in an evaporating dish [1] on a water bath; [1] stop heating when small crystals first appear around the edge of the solution; [1] the rest of the water is then left to evaporate to obtain sodium chloride crystals [1]
7. diagram: three concentric circles with dot or Si at centre, innermost circle with two electrons (dots or crosses); [1] next circle with eight electrons; [1] outer circle with four electrons [1]

8. both have two electrons in their highest energy level (outer shell) [1]
9. $Mg(s) + 2HCl(aq) \rightarrow MgCl_2(aq) + H_2(g)$ [1 for formulae, 1 for balancing, 1 for state symbols]
10. 0.144 nm [1]
11. a isotopes [1]
    b have a different number of neutrons, or ^1_1H has no neutrons, ^2_1H has one neutron and ^3_1H has two neutrons [1]
12. Thomson – tiny negatively charged electrons embedded [1] in a cloud of positive charge [1] Rutherford – the positive charge is concentrated in a nucleus in the centre of the atom; [1] the electrons are orbiting the nucleus [1]

C2.1 Development of the periodic table
1. in order of atomic weight
2. he left gaps for undiscovered elements

C2.2 Electronic structures and the periodic table
1. they have the same number of electrons in the highest occupied energy level or outer shell
2. due to their very stable electron arrangements; they have complete outer shells / energy levels

C2.3 Group 1 – the alkali metals
1. to prevent reaction with air
2. they react with water to produce alkalis or their hydroxides are alkalis
3. potassium bromide, KBr

C2.4 Group 7 – the halogens
1. they have small molecules made of pairs of atoms
2. add chlorine (water) to a solution of potassium bromide (or other soluble bromide); bromine will be displaced

C2.5 Explaining trends
1. when metals react, their atoms lose electrons; as the atoms get larger there are more occupied energy levels and the electrons in the highest occupied level or outer shell are less strongly attracted by the nucleus and so are lost more easily
2. lithium's outer electron is closer to the nucleus and therefore more firmly held

C2.6 The transition elements
1. in the centre block
2. higher melting points, stronger, harder, denser, less reactive (with oxygen and water), ions have different charges, coloured compounds, catalysts

C2 Summary questions
1. in order of their atomic (proton) number; [1] the elements are lined up in groups (vertical columns) with similar properties [1]
2. noble gases [1]
3. they have the same number of electrons in the highest energy level or outer shell (same number of outer electrons) [1]

4. a lithium + water → lithium hydroxide + hydrogen [1 for reactants, 1 for products]
   b three from: lithium floats, moves around the surface, gradually disappears, bubbles (of gas), or fizzes [3]
   c add (universal) indicator; [1] goes purple or blue (or correct alkaline colour for named indicator) [1]
5. sodium reacts faster or melts (lithium does not melt) [1]
6. **physical** three from: high melting point (and/ or boiling point), hard, strong, high density, malleable, ductile, good conductor (of heat and electricity), magnetic [3]
   chemical three from: unreactive or reacts slowly with oxygen (air) and/or water, forms positive ions / ionic compounds, forms ions with different charges, coloured compounds, catalyst [3]
7. they increase [1]
8. a NaBr [1]
   b colourless or white, crystals or solid; [1] dissolves in water; [1] forms a colourless solution [1]
9. a from colourless [1] to orange/yellow/brown [1]
   b chlorine + potassium bromide → potassium chloride + bromine [1 for reactants, 1 for products]
   c $Cl_2 + 2KBr \rightarrow 2KCl + Br_2$ [1 for formulae, 1 for balancing]
10. the elements fitted the groups better or the elements within a group all had similar properties; [1] it allowed for undiscovered elements or left gaps for undiscovered elements [1]
11. $2Fe + 3Cl_2 \rightarrow 2FeCl_3$ [1 for formulae, 1 for balancing]
12. a a sodium atom has more occupied energy levels / shells, [1] so its outer electron is further from [1] and less strongly held by the nucleus, [1] and so can be more easily lost when it reacts [1]
    b a fluorine atom has fewer occupied energy levels / shells, [1] so its nucleus has a greater attraction for electrons [1] in the highest occupied energy level / outer shell, [1] so it attracts electrons more readily when it reacts [1]

C3.1 States of matter
1. solids, liquids, and gases
2. the particles are represented by small solid spheres
3. liquids and gases
4. have different strength forces between particles

C3.2 Atoms into ions
1. compound formed from non-metals
2. KBr, Na_2O, MgO
3. diagram: sodium atom showing one electron, chlorine atom with seven electrons, sodium ion with no electrons and positive charge, chloride ion with eight electrons and negative charge

C3.3 Ionic bonding

1 the oppositely charged ions are held together by strong forces of attraction
2 Ca^{2+}

C3.4 Giant ionic structures

1 they have giant structures with strong electrostatic forces that hold the ions together and a lot of energy is needed to overcome the forces
2 the ions can move freely and carry the charge

C3.5 Covalent bonding

1 Cl–Cl, H–Cl, H–S–H, O=O, O=C=O

C3.6 Structure of simple molecules

1 does not show the true shape of the molecule; all electrons are actually equal but shown differently on dot and cross diagrams; electrons are constantly moving but shown as fixed on diagram
2 intermolecular forces
3 has no overall charge on molecule

C3.7 Giant covalent structures

1 each atom is joined to four other atoms by covalent bonds; so large amount of energy needed to break the many strong bonds
2 similarities: forms of carbon; giant covalent structures or covalent bonding
differences: carbon atoms in diamond are bonded to four other carbon atoms, only to three other atoms in graphite; diamond is three-dimensional, graphite is two-dimensional; diamond is hard, graphite is slippery/soft; diamond is transparent, graphite is grey/opaque; graphite is a good conductor of heat / electricity, diamond is a poor conductor; graphite has delocalised electrons, diamond does not; graphite has intermolecular forces, diamond does not

C3.8 Fullerenes and graphene

1 carbon
2 similarities: forms of carbon covalent bonding, hexagonal rings
differences: graphite has layers, graphene layer only one atom thick

C3.9 Bonding in metals

1 in a giant structure, closely packed together in layers with a regular pattern
2 electrostatic forces between positive (metal) ions and delocalised electrons

C3.10 Giant metallic structures

1 they are harder than pure metals; they can be made or designed to have specific / special properties, such as shape memory alloys
2 delocalised electrons move rapidly through the metal structure

C3.11 Nanoparticles

1 a very small particle that is a few nanometres in size or made of a few hundred atoms

2 5×10^{-9} metres
3 5 nm

C3.12 Applications of nanoparticles

1 used in cosmetics; to deliver drugs in the body, in wound dressings; to give computers vastly improved memory capacities and speeds; as catalysts
2 breathing in tiny particles could damage the lungs; nanoparticles have unpredictable effects on our cells; nanoparticles could affect aquatic life by accumulating in organisms over time

C3 Summary questions

1 giant covalent structure [1]
2 a mixture of metals or a mixture of metal and carbon [1]
3 a outer electrons (electrons in the highest occupied energy level or outer shell) [1]
  b metal atoms lose electrons [1] and non-metal atoms gain electrons [1] or they are transferred from a metal atom to a non-metal atom
  c they are shared [1]
4 in regular patterns [1] in layers [1]
5 a they lose their one [1] outer [1] electron [1] (one electron in the highest occupied energy level or outer shell)
  b they gain one [1] electron [1] so their outer shell has eight electrons [1], or so they have the structure of a noble gas
6 diagram: central N with a lone pair of electrons [1] and three shared pairs of electrons (o x) [1] around it and an H outside each of the pair of electrons
7 it contains delocalised electrons; [1] the delocalised electrons are free to move [1] and carry the electrical charge [1]
8 the particles in a gas have a random arrangement [1] with a great deal of space between them; [1] they move around at high speeds [1] in all directions [1]
9 nanoparticles have a very high surface area to volume ratio [1], which means that smaller quantities are needed [1]
10 it has a giant ionic structure [1] with strong electrostatic forces acting in all directions [1] that need a lot of energy to overcome / break them [1]
11 they have weak intermolecular forces / weak forces between molecules [1] and overcoming these forces does not take much energy [1]
12 a 7×10^{-8} m [1]
   b 200 nm [1]

C4.1 Relative masses and moles

1 it has (two main) isotopes and the relative atomic mass is an average value
2 $23 + 16 + 1 = 40$
3 $\frac{2.2}{44} = 0.05$ mol

C4.2 Equations and calculations

1 2
2 $CaCO_3 = 100$, $CaO = 56$, one mole $CaCO_3$ gives one mole CaO or 100 g $CaCO_3$ gives 56 g CaO, 10 g $CaCO_3$ gives $\left(\frac{10}{100}\right) \times 56 = 5.6$ g CaO

C4.3 From masses to balanced equations

1 ratio $= 0.2 : 0.2 : 0.4$
  simplest ratio $= 1 : 1 : 2$
  equation $H_2 + Cl_2 \rightarrow 2HCl$
2 the reactant that gets used up first in a reaction

C4.4 The yield of a chemical reaction

1 $2Ca + O_2 \rightarrow 2CaO$, 80 g Ca $\rightarrow$ 112 g CaO, 4 g Ca $\rightarrow$ 5.6 g CaO, $\left(\frac{4.4}{5.6}\right) \times 100 = 78.6\%$
2 reactions may not go to completion, other reactions may happen, some product may be lost when separated or collected
3 to help conserve resources, to reduce waste and / or pollution

C4.5 Atom economy

1 the amount of starting materials (the reactants) that end up as the desired product
2 they use fewer natural resources and give fewer waste products, so there will be fewer costs for waste treatment and disposal

C4.6 Expressing concentrations

1 $\frac{2.8}{4} = 0.7$ g/dm^3

C4.7 Titrations

1 pipette

C4.8 Titration calculations

1 0.17 mol/dm^3

C4.9 Volumes of gases

1 $\frac{48}{24} = 2$ moles
2 10 dm^3

C4 Summary questions

1 to show the end point of the reaction or the point at which the acid and alkali have reacted completely [1]
2 $(1 \times 2) + 32 + (16 \times 4)$ [1] $= 98$ [1]
3 to save energy; [1] to conserve limited resources [1]
4 a 14.2 g [1]
  b $\frac{9.20}{23} : \frac{14.2}{71} : \frac{23.4}{58.5}$ [1] $0.4 : 0.2 : 0.4$ [1]
    $2 : 1 : 2$ [1] $2Na + Cl_2 \rightarrow 2NaCl$ [1]
5 $\frac{8.0}{40}$ [1] $= 0.2$ moles [1]
6 there is only one product / or all the atoms in the reactants appear in the product [1]
7 100 g $CaCO_3$ produces 56 g CaO [1]
  150 g $CaCO_3$ produces 84 g CaO [1]
  so % yield $= \frac{67.2}{84} \times 100$ [1] $= 80\%$ [1]

Answers

8   a   4.48 g [1]

    b M_r KOH = 56, [1] so moles = $\frac{4.48}{56}$ [1]

        = 0.08 mol/dm³ [1]

9   a   400 cm³ [1]

    b   400 cm³ [1]

10   moles KOH = $\frac{0.10}{1000} \times 25$ = 0.0025 [1]

    so moles NaOH = 0.0025 [1]

    concentration of HCl = $\frac{0.0025}{20} \times 1000$ [1]

        = 0.125 mol/dm³ (or 0.13 mol/dm³ to 2 s.f.) [1]

Section 1 Practice questions

01.1   11 [1]

01.2   23 [1]

01.3   11 [1]

01.4   12 [1]

01.5   2,8,1 [1]

01.6   alkali metals, [1] Group 1 [1]

01.7   sodium + chlorine → sodium chloride [1]

01.8 $2Na(s) + Cl_2(g) \rightarrow 2NaCl(s)$ [1 for balancing, 1 for state symbols]

02.1   Li: metallic; [1] F_2: covalent; [1] LiF: ionic [1]

02.2 F_2 [1]

02.3   Li [1]

02.4   Li and LiF [2]

02.5   F at centre of two concentric circles with two crosses on inner circle and seven crosses on outer circle [1]

02.6   Li at centre of one circle with two crosses [1] with brackets and + sign top right outside bracket; [1] F at centre of two concentric circles with two crosses on inner circle and eight crosses on outer circle [1] enclosed in brackets with – sign top right outside brackets [1]

02.7   a particle that is a few nm (nanometres) in size **or** containing a few hundred atoms [1]

02.8   much larger surface area **or** react much faster [1]

03.1   in order of atomic number [1]

03.2   potassium and argon; [1] tellurium and iodine [1]

03.3   they are the same [1]

03.4   the number of electrons in the highest occupied energy level is the same as the group number [1]

03.5   relative atomic masses depend on the number of isotopes [1] and the number of neutrons in the atom [1] but group number depends on the number of electrons [1]

03.6   protons / electrons had not been discovered [1]

04.1   giant structure, [1] covalent [1] strong bonds, [1] each atom bonds to four other atoms [1]

04.2   covalent / strong bonds [2]

04.3   (buckminsterfullerene has) one molecule containing 60 carbon atoms, [1] hexagonal rings, [1] hollow shape [1]

04.4   two from: drug delivery, lubricants, catalysts, reinforcing composite materials, electronics industry [2]

05.1   24 + (35.5 × 2) [1] = 95 [1]

05.2 A_r Mg = 24, M_r MgCl$_2$ = 95 [1]

    mass of MgCl$_2$ = $\frac{95 \times 6}{24}$ [1] = 23.75 g [1]

05.3   contains positive and negative ions [1] that are free to move to carry charge [1]

05.4   doesn't show shape **or** doesn't reflect total number of ions bonded [1]

06.1   measure 25 cm³ sodium hydroxide into a conical flask [1] using a pipette; [1] add indicator to conical flask; [1] add acid from burette until indicator changes colour; [1] continue until indicator changes colour permanently; [1] record volume of acid added [1]

06.2   15.55 cm³ [1]

06.3 $\frac{14.95 + 15.05 + 15.00}{3}$ [1] = 15.00 cm³ [1]

06.4   moles NaOH = $\frac{0.100 \times 25}{1000}$ = 0.0025 [1]

    moles H$_2$SO$_4$ = 0.00125 [1]

    concentration of H$_2$SO$_4$ = $\frac{0.00125 \times 1000}{15.00}$

        = 0.083 [1] moles per dm³ [1]

06.5 $\frac{0.6 \times 1000}{25}$ [1] = 24 g/dm³ [1]

C5.1 The reactivity series

1   rock from which metal can be extracted economically

2   (most reactive) potassium, magnesium, copper (least reactive)

C5.2 Displacement reactions

1   zinc + copper sulfate → zinc sulfate + copper
    $Zn(s) + CuSO_4(aq) \rightarrow ZnSO_4(aq) + Cu(s)$

2   magnesium, carbon, iron, hydrogen, copper

3   magnesium is oxidised; copper ions are reduced

C5.3 Extracting metals

1   gold is unreactive

2   a   two from: zinc, iron, tin, lead, copper

    b   reduction

C5.4 Salts from metals

1   any metal that is more reactive than hydrogen, but less reactive than calcium (e.g., lead, tin, iron, aluminium, magnesium)

2   magnesium chloride

C5.5 Salts from insoluble bases

1   to ensure all the acid has reacted

C5.6 Making more salts

1   water

2   magnesium carbonate + hydrochloric acid → magnesium chloride + water + carbon dioxide
    $MgCO_3(s) + 2HCl(aq) \rightarrow MgCl_2(aq) + H_2O(l) + CO_2(g)$

C5.7 Neutralisation and the pH scale

1   hydrogen ions, $H^+(aq)$

2   a soluble base or a substance that produces hydroxide ions in solution, $OH^-(aq)$

3   universal indicator or full-range indicators

C5.8 Strong and weak acids

1    $CH_3COOH(aq) \rightleftharpoons CH_3COO^-(aq) + H^+(aq)$

2   1×10^{-4} mol/dm³

3   1×10^{-5} mol/dm³ or 0.00001 mol/dm³

C5 Summary questions

1   ore [1]

2   gold is very unreactive [1]

3   electrolysis [1]

4   a   alkali [1]

    b   neutralisation [1]

    c   add an indicator; [1] will change colour at the endpoint [1]

    d   sodium hydroxide + hydrochloric acid → sodium chloride + water [1]

5   hydrogen ion or $H^+(aq)$ [1]

6   zinc chloride [1] and hydrogen [1]

7   addition of oxygen or loss of electrons [1]

8   a   reduction / redox [1]

    b   lead oxide + carbon → lead + carbon dioxide [1]

9   add copper oxide to sulfuric acid, until there is an excess; [1] gently warm the mixture; [1] filter the solution; [1] evaporate the water [1] until crystals of copper sulfate start to form; [1] leave to cool and crystallise [1]

10   strong acids completely ionise in water; [1] weak acids only partly ionise in water [1]

11    $Mg(s) + CuSO_4(aq) \rightarrow MgSO_4(aq) + Cu(s)$ [1 for formulae, 1 for state symbols]

12   a   a magnesium atom loses two electrons [1] to form a magnesium ion [1]

    b   oxidation [1]

13   0.01 mol/dm³ [1]

C6.1 Introduction to electrolysis

1   they must be melted or dissolved in water

2   a   chlorine

    b   zinc

C6.2 Changes at the electrodes

1   reduction / positive sodium ions gain electrons

2    $2Cl^- \rightarrow Cl_2 + 2e^-$

3   a   oxygen

    b   hydrogen

C6.3 The extraction of aluminium

1   to lower the melting temperature of the mixture

2   aluminium and oxygen; oxygen then reacts with the electrode to produce carbon dioxide

C6.4 Electrolysis of aqueous solutions

1   the solution contains hydrogen ions that are discharged in preference to sodium ions because sodium is more reactive than hydrogen

2   hydrogen, chlorine, potassium hydroxide

C6 Summary questions

1   anode [1]

2   positive / metal [1]

3 a positive electrode: chlorine; [1] negative electrode: calcium [1]
   b positive electrode: iodine; [1] negative electrode: lithium [1]
   c positive electrode: bromine; [1] negative electrode: magnesium [1]
4 ionic [1]
5 hydrogen / H^+(aq) ions [1] and hydroxide / OH^-(aq) ions [1]
6 a large amounts of energy [1] and high temperatures required [1]
   b aluminium oxide [1]
   c to lower the melting point of the mixture [1]
   d made of carbon/graphite, [1] which reacts with the oxygen [1] produced to form carbon dioxide, [1] so electrode wears away [1]
7 a hydrogen, [1] chlorine, [1] and sodium hydroxide [1]
   b add sodium chloride solution [1] to an electrolysis cell; [1] pass a current through; [1] collect gases produced; [1] test the gases collected; [1] test the solution with an indicator [1]
8 negative electrode: sodium ions are attracted [1] and gain electrons or are reduced [1] to form sodium atoms / metal; [1] positive electrode: chloride ions are attracted [1] and lose electrons or are oxidised [1] to chlorine atoms / molecules / gas [1]
9 negative electrode: $Pb^{2+} + 2e^- \rightarrow Pb$; [1] positive electrode: $2Br^- \rightarrow Br_2 + 2e^-$ [1]
10 a both the positively charged potassium / K^+ ions [1] and hydrogen / H^+ ions [1] (from water) are attracted to the negative electrode; the less reactive element (hydrogen) will be formed [1]
   b negative electrode: $2H^+$(aq) $+ 2e^- \rightarrow H_2$(g); [1] positive electrode: $4OH^-$(aq) $\rightarrow 2H_2O$(l) $+ O_2$(g) $+ 4e^-$ [1]

C7.1 Exothermic and endothermic reactions

1 it transfers energy to the surroundings or heats the surroundings
2 either it cools the surroundings or it needs to be heated to keep it going
3 repeat the investigation and take a mean; insulate the container; use a lid and a temperature sensor

C7.2 Using energy transfers from reactions

1 **advantage** one from: less waste, less materials / resources used
   disadvantage one from: has to be heated or needs energy so it can be used again, slower reaction, smaller temperature rise
2 **advantage** one from: can be used anywhere, can be stored easily (ice needs to be made and / or stored in special equipment)

disadvantage one from: can only be used once, more waste, possibly more hazardous than ice

C7.3 Reaction profiles

1 products are at a lower energy level
2 energy level diagram: similar to figure but with the products at a higher level than the reactants, the reaction pathway rising above the products level, and the activation energy labelled with an arrow pointing upwards from the reactants level to the top of the pathway curve

C7.4 Bond energy calculations

1 bonds broken: $(2 \times C–C) + (8 \times C–H) + (5 \times O=O)$
   energy needed = 6488 kJ/mol
   bonds made: $6 \times C=O + 8 \times H–O$ = 8542 kJ/mol
   energy change of reaction = 2054 kJ/mol

C7.5 Chemical cells and batteries

1 a battery is made up of two or more cells
2 from zinc atoms to copper ions

C7.6 Fuel cells

1 no pollutants are produced
2 water

C7 Summary questions

1 endothermic [1]
2 do not need to be electrically recharged, [1] no pollutants are produced [1]
3 sports injury pack or to cool drinks [1]
4 two from: combustion, oxidation, neutralisation [2]
5 the minimum amount of energy needed [1] before colliding particles of reactants have sufficient energy to cause a reaction [1]
6 a join two metals together [1] by a wire [1] and dip them into an electrolyte or salt solution [1]
   b choose metals with a greater difference in reactivity [1]
7 $2H_2 + O_2 \rightarrow 2H_2O$ [1 for formulae, 1 for balancing]
8 the energy level of the reactants is below the level of the products [1]
9 this is an endothermic reaction; [1] calcium carbonate needs to absorb energy [1] from the surroundings to be broken down
10 the greater its tendency to form a positive ion, [1] the more reactive the metal is [1]
11 reaction profile diagram: reactants on line above products; [1] reaction pathway shown as curved line reaching maximum above reactants; [1] activation energy shown as an arrow from reactants to maximum of pathway; [1] energy change of reaction shown as arrow pointing downwards from reactants to products [1]
12 bonds broken: $(1 \times C=C) + (4 \times C–H) + (3 \times O=O)$ [1] energy needed = 3758 kJ/mol [1]
   bonds made: $(4 \times C=O) + (4 \times H–O)$ [1] = 5076 kJ/mol [1]
   energy change of reaction = 1318 kJ/mol [1]

reaction is exothermic / energy is transferred to the surroundings [1]

Section 2 Practice questions

01.1 exothermic [1]
01.2 a substance that speeds up the reaction [1] without being used up in the reaction [1]
01.3 two from: increase the surface area of the iron, use smaller particles of iron, increase the concentration of salt, shake the pack [2]
02.1 hydrogen [1]
02.2 $\frac{44}{0.72}$ [1] 61 [1] kJ [1]
02.3 hydrogen has a much greater volume as is a gas, [1] ethanol is a liquid [1]
02.4 two from: produces a very large amount of energy per gram, only product is water / forms non-polluting product; hydrogen is very flammable [2]
02.5 two from: need modified engines, hydrogen is difficult to store, few places to buy the fuel [2]
03.1 CuO(s) $+ 2HCl$(aq) $\rightarrow CuCl_2$(aq) $+ H_2O$(l) [1 for H_2O, 1 for correct balancing, 1 for state symbols]
03.2 a base [1]
03.3 place hydrochloric acid in a suitable container; [1] add copper oxide; [1] stir / gently warm; [1] until copper oxide no longer dissolves / black solid remains [1]
03.4 to ensure that all acid reacts [1]
03.5 filter to remove excess copper oxide; [1] evaporate water; [1] suitable method to ensure that only some of solution evaporates; [1] leave to crystallise [1]
04.1 H_2SO_4(aq) $+ 2NaOH$(aq) $\rightarrow Na_2SO_4$(aq) $+ 2H_2O$(l) [3 – 1 for correct formula, 1 for balancing, 1 for state symbols]
04.2 measure 25 cm³ of sulfuric acid; [1] add to polystyrene cup; [1] measure temperature of sulfuric acid; [1] add 5 cm³ of sodium hydroxide solution to sulfuric acid; [1] measure highest temperature reached by the mixture; [1] repeat using different volumes of sodium hydroxide [1]
04.3 exothermic [1]
04.4 as volume of sodium hydroxide increases, temperature increases until 25 cm³ is added [1] because energy is being given out by the reaction; [1] after 25 cm³, temperature decreases when volume of sodium hydroxide increases [1] because reaction has finished or all acid is used up [1]
05.1 chlorine [1]
05.2 hydrogen [1]
05.3 sodium hydroxide [1]
05.4 sodium is more reactive than hydrogen, [1] so hydrogen discharged in preference to sodium [1]
05.5 $2Cl^-$(aq) $\rightarrow Cl_2$(g) $+ 2e^-$ [1 for symbols and formulae, 1 for balancing, 1 for state symbols]
05.6 oxidation [1] because chloride ion loses electrons [1]

Answers

06.1 bonds broken: $4 \times C–C + 12 \times C–H + 8 \times O=O$
= 10 328 kJ/mol [2]
bonds made: $10 \times C=O + 12 \times H–O$
= 13 618 kJ/mol [2]
energy change = 3290 kJ/mol [1]

06.2 energy released when bonds made (in products) greater [1] than energy needed to break bonds (in reactants) [1]

C8.1 Rate of reaction
1 amount (of reactant or product) and time
2 the line decrease in value before gradually levelling off

C8.2 Collision theory and surface area
1 activation energy
2 powders have greater surface area than large lumps of solid, and this increases the chance of collisions

C8.3 The effect of temperature
1 it increases the energy of the particles and the frequency of collisions

C8.4 The effect of concentration and pressure
1 it increases the frequency of collisions
2 the frequency of collisions increases because there are more particles in the same volume

C8.5 The effect of catalysts
1 they are not used up in the reaction
2 they reduce the energy needed and the time needed, reducing costs; they may reduce the amount of fossil fuel used, conserving resources and reducing pollution

C8.6 Reversible reactions
1 a reaction that can go both forwards and backwards / both ways / in both directions

C8.7 Energy and reversible reactions
1 it is an endothermic reaction
2 it is an exothermic reaction

C8.8 Dynamic equilibrium
1 when the rates of the forward and reverse reactions of a reversible reaction are equal or when the amounts of reactants and products in a reversible reaction are constant
2 so that more of the reactants react or so that more SO_3 is produced

C8.9 Altering conditions
1 an increase in pressure
2 a decrease in temperature

C8 Summary questions
1 reversible reaction [1]
2 fastest: iron filings; [1] slowest: a block of iron [1]
3 a substance that alters the rate of a reaction [1] without itself being used up [1]

4 gas syringe [1] and inverted measuring cylinder filled with water [1]
5 a a suspension [1] of an insoluble solid [1]
   b place a conical flask [1] on top of a cross drawn on a piece of paper; [1] place solutions in the conical flask; [1] start a stopwatch; [1] note when the cross can no longer be seen; [1] record the time taken [1]
6 measuring the decreasing mass of the reaction mixture; [1] measuring the increasing volume of gas given off [1]
7 draw a tangent to the curve; [1] find the gradient of the tangent [1]
8 particles are closer together / more particles in the same volume; [1] so collisions are more frequent or more collisions per second [1]
9 particles collide more frequently [1] and with more energy, [1] so more collisions have the activation energy or minimum energy needed for reaction [1]
10 the equilibrium will move to the left / the backward reaction will increase; [1] more ICl will be produced [1]
11 there are three molecules of reactant gases and two molecules of product gases / fewer molecules of gas in products than reactants; [1] so high pressure favours forward reaction [1] and more sulfur trioxide is produced [1]
12 by increasing the temperature [1]

C9.1 Hydrocarbons
1 liquids with different boiling ranges separated from a mixture of liquids (crude oil)
2 it has the general formula C_nH_{2n+2}; it is a saturated compound; it has only single carbon–carbon bonds
3 C_4H_{10}

C9.2 Fractional distillation of oil
1 different hydrocarbons have different boiling points (so they condense at different levels)

C9.3 Burning hydrocarbon fuels
1 ethane + oxygen → carbon dioxide + water
2 carbon monoxide, carbon, unburnt hydrocarbons, water

C9.4 Cracking hydrocarbons
1 one from: to make fuels that are more useful or for which there is more demand; large hydrocarbons do not burn easily or are less in demand
2 three from: are unsaturated; have a double bond; have a different general formula; have fewer hydrogen atoms than the corresponding alkane; are more reactive; react with or decolourise bromine water

C9 Summary questions
1 to make fuels that are more useful [1]
2 carbon dioxide [1] and water [1]
3 carbon dioxide is a greenhouse gas [1] and may cause global warming [1]

4 two from: to make alkenes; to make alkanes with smaller molecules; to make fuels that are more useful or for which there is more demand; to make polymers (from alkenes) [2]
5 by heating a mixture of hydrocarbon vapours and steam [1] to a very high temperature, [1] or by passing hydrocarbon vapours [1] over a hot catalyst [1]
6 $C_{12}H_{26} \rightarrow C_6H_{14} + C_4H_8 + C_2H_4$ [1]
7 ignite / burn more easily, [1] thin / runny liquids, [1] burn with a clean flame / little smoke [1]
8 carbon monoxide, [1] which is a poisonous gas, [1] may be produced in a limited supply of air [1]
9 a 5C joined to 12H, all single bonds, each C with four single bonds ($CH_3CH_2CH_2CH_2CH_3$) [1]
   b four from: alkane; saturated; hydrocarbon; general formula C_nH_{2n+2}; has only single bonds; burns to produce carbon dioxide and water [4]
10 19 [1]
11 $2C_4H_{10} + 13O_2 \rightarrow 8CO_2 + 10H_2O$ [1 for formulae, 1 for balancing]
12 crude oil vapour [1] enters column; vapour rises [1] until boiling point of compound is reached; [1] vapour condenses (at that level); [1] high-boiling fractions collected at bottom of column, low-boiling fractions collected at top [1]

C10.1 Reactions of the alkenes
1 C_6H_{12}

C10.2 Structures of alcohols, carboxylic acids, and esters
1 $CH_3CH_2CH_2OH$
2

C10.3 Reactions and uses of alcohols
1 it is a solvent (it mixes with water); it evaporates easily
2 one from: apply a flame – ethanol burns in air; add sodium – ethanol reacts more slowly (also sodium sinks in ethanol); add (acidified) potassium dichromate (or other oxidising agent) and heat – colour change or smell of vinegar for ethanol

C10.4 Carboxylic acids and esters
1 they produce $H^+(aq)$ / hydrogen ions in aqueous solutions
2 they have distinctive smells / tastes that are fruity / pleasant
3 $CH_3CH_2COOH(aq) \rightleftharpoons CH_3CH_2COO^-(aq) + H^+(aq)$

C10 Summary questions
1 C_3H_6; [1], C_4H_8; [1] C_8H_{16} [1]
2 due to the carbon–carbon double bond [1]
3 carbon dioxide [1]
4 propanoic acid, [1] CH_3CH_2COOH [1]
5 methanol, [1] CH_3OH; [1] ethanol, [1] CH_3CH_2OH; [1] propanol, [1] $CH_3CH_2CH_2OH$ [1]

6 fizzing / effervescence; [1] produces hydrogen [1] (slower than with water)

7 microbes in the air [1] caused oxidation of ethanol, [1] producing ethanoic acid [1]

8 **a** ethyl ethanoate [1]

b (sulfuric) acid catalyst, [1] heat the mixture [1]

9 **a** glucose is fermented [1] with yeast [1]

b ethene is reacted with steam [1] in the presence of a catalyst [1]

10 they burn with a smoky, yellow (luminous) flame [1] because there is incomplete combustion; [1] they release less energy per mole than alkanes [1]

11 $2CH_3CH_2CH_2OH + 9O_2 \rightarrow 6CO_2 + 8H_2O$ [1 for formulae, 1 for balancing]

12 test with universal indicator or pH meter, [1] ethanoic acid has higher pH; [1] **or** add a metal carbonate, ethanoic acid reacts more slowly

13 hydrochloric acid ionises completely in water; [1] ethanoic acid only partly ionises in water [1]

C11.1 Addition polymerisation

1 from many small molecules or monomers that join together or polymerise to make a large or very long molecule

2 thousands or a very large number

3 alkenes are more reactive / are unsaturated / have a double bond; alkanes are unreactive / unsaturated / do not have a double bond

C11.2 Condensation polymerisation

1 two

2 a diol and a dicarboxylic acid

C11.3 Natural polymers

1 starch and cellulose

2 two different functional groups join together to produce the polymer and water

C11.4 DNA

1 nucleotides

2 four

C11 Summary questions

1 two from: plastic carrier bags, drinks bottles, dustbins, washing-up bowls, cling film [2]

2 **a** double helix, [1]

b two polymer chains [1]

3 **a** propene, [1]

b C_3H_6 [1]

4 **a** natural polymers are found in all living things, [1] synthetic polymers are manufactured [1]

b starch made from sugars / glucose; [1] cellulose made from sugars / glucose; [1] proteins made from amino acids [1]

5 **a** four

b DNA monomers are called nucleotides; [1] lots of nucleotides join together to form a polynucleotide, DNA [1]

6 an addition polymer has one product, [1] a condensation polymer has two [1]

7 diol + dicarboxylic acid → polyester + water [1]

8 a polymer has very large molecules [1] made from many small molecules / monomers, [1] joined [1] together by a polymerisation reaction [1]

9 in addition polymerisation the monomers are alkenes / contain a double bond, [1] in condensation polymerisation the monomers have two different functional groups; [1] in addition polymerisation one type of monomer is used, [1] in condensation polymerisation two different monomers are used [1]

10

[2]

11 two from: they are made using different reaction conditions; they have different structures; they have different-shaped molecules [2]

12 amino acids have two functional groups, [1] one basic (the amine group, $-NH_2$) [1] and one acidic (the carboxylic acid group, $-COOH$); [1] the acid and base groups react together [1] to form a condensation polymer [1]

Section 3 Practice questions

01.1 carbon, hydrogen, oxygen [2 if all correct, 1 if one or two correct]

01.2 18 [1]

01.3 starch, [1] cellulose [1]

02.1 alcohols [1]

02.2 C_3H_7OH [1]

02.3 $-OH$ [1]

03.1 reversible [1]

03.2 endothermic [1]

03.3 add water [1]

03.4 energy will be given out [1]

03.5 the temperature would increase [1]

03.6 add anhydrous copper sulfate to the liquid [1]

03.7 blue powder / crystals would be formed if water was present [1]

04.1 contains hydrogen and carbon [1] only [1]

04.2 carbon–carbon double bond [1]

04.3 add bromine water, [1] solution turns colourless if a carbon–carbon double bond is present [1]

04.4 $C_2H_4 + 3O_2 \rightarrow 2CO_2 + 2H_2O$ [1 for CO_2, 1 for H_2O, 1 for correct balancing]

04.5 incomplete combustion, [1] so carbon / soot particles formed [1]

04.6 poly(ethene) / polythene [1]

04.7 many ethene molecules / monomers [1] join together [1] to form a very large molecule / long chain [1]

05.1 sensible scales using at least half the grid for the points, [1] both axes correctly labelled (time in seconds and volume of oxygen in cm^3), [1] all points correct [2 – 1 if eight/nine correct], suitable best fit line drawn [1]

05.2 5 mins / 65 cm^3 [1]

05.3 slope or gradient was steepest at the start [1]

05.4 two from: concentration of reactants highest at the start / concentration of reactants decreases with time, more (reactant / hydrogen peroxide) particles at start, greater frequency of collisions at start / more collisions per second at start [2]

05.5 draw a tangent to the curve at 40 seconds, [1] find the gradient of the tangent [1]

05.6 line with steeper slope initially, [1] levelling off at same volume as other line [1]

05.7 two from: same volume of hydrogen peroxide, same concentration of hydrogen peroxide, same mass of manganese(IV) oxide [2]

06.1 $C_{14}H_{30}$ [1]

06.2 high temperature, [1] catalyst or steam [1]

06.3 displayed formula for propene: three carbon atoms, with one single carbon–carbon bond and one double carbon–carbon bond; [1] six hydrogen atoms, each with a single bond [1] to a carbon atom so that carbon atoms all have a total of four bonds [1]

06.4 C_4H_8 [1]

06.5 poly(propene) / poly(alkenes) / poly(ethene): are non-biodegradable / cannot be broken down by microorganisms (in soil/ environment) [1] and last for a very long time (hundreds or more years); [1] are bulky and are used in large amounts as disposable packaging, becoming litter and taking up space in landfill; [1] are made from crude oil / use up finite resources; [1] are difficult to identify and / or separate from waste, making recycling difficult; [1] incineration produces carbon dioxide (leading to global warming) [1]

C12.1 Pure substances and mixtures

1 a pure substance is one that is made up of just one substance

2 a mixture

C12.2 Analysing chromatograms

1 mobile phase and stationary phase

C12.3 Testing for gases

1 oxygen

C12.4 Tests for positive ions

1 calcium (Ca^{2+}) and magnesium (Mg^{2+})

C12.5 Tests for negative ions

1 carbon dioxide

2 hydrochloric acid contains chloride ions (Cl^-) and sulfuric acid contains sulfate ions (SO_4^{2-}), which both give precipitates with silver nitrate solution

C12.6 Instrumental analysis

1 two from: accurate, quicker, sensitive, very small samples can be analysed

2 comparing with a database

Answers

C12 Summary questions

1 (artificial) colourings [1]
2 green [1]
3 add dilute hydrochloric acid, [1] followed by barium chloride solution; [1] if sulfate ions are present, a white precipitate is formed [1]
4 in chemistry, pure means made up of just one substance; [1] orange juice is made up of more than one substance [1]
5 paint is a mixture of substances [1] in set proportions [1] so that the formulation has the desired properties [1]
6 **advantages** two from: highly accurate; sensitive; very fast; enable very small samples to be analysed [2]
   disadvantages two from: very expensive; need special training to use; results need to be compared with a database [2]
7 a flame emission spectroscopy [1]
   b the metal ions give different colours, so colours of metal ions are masked [1] and you wouldn't be able to identify them [1]
8 $R_f = \dfrac{5}{20}$ [1] = 0.25 [1]
9 copper(II) carbonate [1 for copper, 1 for carbonate]
10 potassium iodide [1 for potassium, 1 for iodide]
11 Fe^{2+} / iron(II) ions [1] and SO_4^{2-} / sulfate ions [1]
12 $Fe^{3+}(aq) + 3OH^-(aq) \rightarrow Fe(OH)_3(s)$ [1 for formulae, 1 for balancing, 1 for state symbols]

C13.1 History of our atmosphere

1 volcanoes
2 photosynthesis

C13.2 Our evolving atmosphere

1 three from: photosynthesis; dissolved in oceans / water; locked up in fossil fuels; locked up in sedimentary rocks (including limestone)
2 78% nitrogen (almost 80%) and 21% oxygen (just over 20%)

C13.3 Greenhouse gases

1 carbon dioxide, methane, water vapour
2 burning of fossil fuels

C13.4 Global climate change

1 rapid increase of the levels of greenhouse gases
2 biofuels are made from plant material that absorbs carbon dioxide during photosynthesis, and returns the carbon dioxide to the atmosphere when it is burnt

C13.5 Atmospheric pollutants

1 carbon monoxide, carbon, unburnt hydrocarbons, water
2 acid rain

C13 Summary questions

1 four from: nitrogen, oxygen, water vapour, carbon dioxide, argon, neon, krypton, xenon [4]
2 carbon dioxide [1]
3 carbon dioxide CO_2, methane CH_4, water vapour H_2O [1 for each correct name, 1 for each correct formula]
4 photosynthesis [1]
5 infrared or longer wavelength radiation [1]
6 burning fossil fuels, [1] decomposition of waste at landfill sites [1]
7 use less electricity, [1] walk or cycle rather than drive cars, [1] recycle waste [1]
8 a rising sea levels; more frequent and severe storms; changes in temperature; changes in the amount, timing and distribution of rainfall; changes to the distribution of wildlife species [4]
   b complex system [1] with many different variables [1]
9 in a limited supply of air, incomplete combustion takes place [1] and carbon monoxide is produced; [1] carbon monoxide is a toxic gas [1]
10 calcium carbonate [1] $CaCO_3$ [1]
11 sulfur burns in air to form sulfur dioxide, [1] which dissolves in rain to form acid rain; [1] acid rain kills animal and plant life in lakes, [1] so by removing sulfur there will be less acid rain, [1] so animals and plants in lakes will not be harmed [1]
12 carbon dioxide + water → glucose + oxygen [1] $6CO_2 + 6H_2O \rightarrow C_6H_{12}O_6 + 6O_2$ [1 for formulae, 1 for balancing]
13 $C_3H_8 + 5O_2 \rightarrow 3CO_2 + 4H_2O$ [1 for formulae, 1 for balancing]
14 the skeletons and shells of the marine organisms [1] built up at the bottom of oceans; [1] they were covered with layers of sediment; [1] under the pressure caused by being buried by the layers of sediment; [1] the deposits formed limestone [1]

C14.1 Finite and renewable resources

1 two from: wool, cotton, wood, rubber, or any other natural resource
2 using a renewable resource; new trees can be planted to provide the fuel

C14.2 Water safe to drink

1 sedimentation and / or filtration to remove solids; killing microbes (disinfecting / sterilising) using chlorine (or other methods, e.g., ozone, ultraviolet)
2 by measuring the boiling point, which should be 100 °C

C14.3 Treating waste water

1 waste water from homes, businesses, and industry
2 to break down organic matter and harmful microorganisms

C14.4 Extracting metals from ores

1 economical method of extracting copper from low-grade ores **or** reduce use of fossil fuels for smelting
2 Iron is more reactive than copper.

C14.5 Life Cycle Assessments

1 to prevent bias or to check the data and the validity of conclusions drawn

C14.6 Reduce, reuse, and recycle

1 to save: the energy needed to extract its ore; resources; fossil fuels; land needed for mining and / or landfill; **or** other specified environmental impact

C14 Summary questions

1 a finite resources are used up at a faster rate than they can be replaced; [1] renewable resources can be replaced at the same rate at which they are used up [1]
   b finite resources: crude oil, [1] coal, [1] gold [1]; renewable resources: sugar cane, [1] cotton, [1] wood [1]
2 to save energy, [1] to conserve metal ores, [1] to reduce pollution [1]
3 to sterilise water [1] by killing microorganisms [1]
4 to trap large solid objects [1]
5 fresh water is rainwater that is found in rivers and lakes; [1] potable water is water that is fit to drink [1]
6 bottled water contains other dissolved substances; [1] pure water only contains the compound water [1]
7 a the solid sediments [1] that have settled out from sewage [1]
   b as a fertiliser, [1] as a source of renewable energy [1]
8 a phytomining, [1] bioleaching [1]
   b conserves copper-rich ores, [1] less impact on the environment [1]
9 a distillation [1]
   b test the boiling point; [1] it should be 100 °C [1]
   c the boiling point would be above 100 °C **or** a boiling point range is obtained [1]
10 a it is used to assess the impact on the environment of products, processes, or services [1]
   b raw material extraction [1] → manufacture [1] → use/reuse/maintenance [1] → recycle / waste management [1]
   c to prevent bias [1]
11 a copper sulfide + oxygen → copper + sulfur dioxide [1]
   b $Cu_2S + O_2 \rightarrow 2Cu + SO_2$ [1 for formulae, 1 for balancing]
   c acid rain [1]
12 a $Fe(s) + CuSO_4(aq) \rightarrow Cu(s) + FeSO_4(aq)$ [1 for formulae, 1 for balancing, 1 for state symbols]
   b $Fe(s) + Cu^{2+}(aq) \rightarrow Cu(s) + Fe^{2+}(aq)$ [1 for formulae, 1 for state symbols]

C15.1 Rusting

1 presence of air and water
2 iron is covered with a layer of a more reactive metal, zinc; water or oxygen react with the zinc rather than iron

C15.2 Useful alloys

1 they are harder than pure metals; they can be made / designed to have specific or special properties
2 50%

C15.3 The properties of polymers

1 they are made using different reaction conditions; they have different structures or different-shaped molecules

2 thermosoftening polymers have no cross-links or no covalent bonds between the polymer chains; thermosetting polymers have cross-links

C15.4 Glass, ceramics, and composites

1 sand and boron trioxide

2 more resistant to bending forces

C15.5 Making ammonia – the Haber process

1 nitrogen + hydrogen → ammonia

2 unreacted gases are recycled

C15.6 The economics of the Haber process

1 the reaction would be too slow (rate decreased and catalyst will not work)

C15.7 Making fertilisers in the lab

1 ammonia + sulfuric acid → ammonium sulfate

2 titration

C15.8 Making fertilisers in industry

1 nitrogen, phosphorus, and potassium

2 can make a fertiliser with specific properties

C15 Summary questions

1 iron, [1] water, [1] air / oxygen [1]

2 aluminium has a low density [1] and is strong [1]

3 a sand, [1] limestone, [1] sodium carbonate [1]
   b sand [1]

4 a i ammonium nitrate, NH_4NO_3,
     ii ammonium sulfate, $(NH_4)_2SO_4$,
     iii ammonium phosphate, $(NH_4)_3PO_4$
        [1 for each name, 1 for each formula]
   b neutralisation [1]

5 a air (nitrogen) [1] and natural gas (hydrogen) [1]
   b 200 atm, [1] 450 °C, [1] iron catalyst [1]
   c the gases are cooled [1] and ammonia condenses; [1] nitrogen and hydrogen remain as gases [1]

6 a N nitrogen, [1] P phosphorous, [1] K potassium [1]
   b NPK fertiliser is a mixture [1] designed as a useful product, [1] made in set proportions [1] so fertiliser has the desired properties [1]

7 a ethene [1]
   b made using different conditions [1] and have different structures [1]

8 ammonia + nitric acid → ammonium nitrate [1]
   NH_3 + HNO_3 → NH_4NO_3 [1]

9 a less reactive metal is coated with a more reactive metal; [1] the more reactive metal reacts with any water or oxygen present [1] rather than the less reactive metal [1]

10 thermosetting polymers do not soften or melt when they get hot; [1] they are good insulators of heat; [1] they can be moulded into shape but are then rigid / hard [1]

11 a lower temperature would increase the yield of ammonia; [1] however, at a low temperature the rate of the reaction would be very slow, [1] so a compromise temperature [1] is used so ammonia is obtained more quickly, but with a lower yield [1]

Section 4 Practice questions

01.1 reversible [1]

01.2 the air [1]

01.3 as a catalyst or to speed up the reaction [1]

01.4 recycled or returned to reactor [1]

01.5 cooled or temperature decreased [1] to below boiling point of ammonia [1]

02.1 yellow [1]

02.2 lithium [1] red [1] **or** potassium [1] lilac [1]

02.3 the amount of radiation depends upon the amount of element in solution [1]

02.4 each element produces a unique pattern (of radiation) [1]

02.5 three from: very small sample required, rapid, sensitive, accurate [3]

03.1 calcium carbonate + calcium oxide → carbon dioxide [1]

03.2 thermal decomposition [1]

03.3 alkali(ne) [1]

03.4 forms calcium carbonate [1], which is a white solid / precipitate **or** insoluble [1]

04.1 fuel [1]

04.2 causes acid rain **or** asthma / respiratory problems [1]

04.3 electrolysis [1]

05.1 to provide enzymes / to speed up reactions / to make copper sulfate or soluble copper compounds [1]

05.2 iron + copper sulfate → iron sulfate + copper [1]

05.3 displacement [1]

05.4 two from: it is acidic or contains sulfuric acid, it contains iron salts / compounds, it may contain other metal compounds, it is harmful / toxic / damaging to plants / animals / humans [2]

06.1 carbon dioxide [1]

06.2 water vapour / methane / ammonia / nitrogen (not oxygen) [1]

06.3 algae and plants [1] by photosynthesis [1]

06.4 there is insufficient evidence / no proof [1]

06.5 fractional distillation [1]

06.6 the liquids are warmed; [1] nitrogen has the lowest boiling point [1] so forms a gas and separates first; [1] argon is next to separate, [1] leaving oxygen [1]

07.1 $((14 + 3) \times 3) + 31 + (16 \times 4)$ [1] = 146 [1]

07.2 isotopes are atoms of the same element **or** atoms with the same number of protons [1] but a different number of neutrons; [1] $^{31}_{15}P$ has 16 neutrons, $^{32}_{15}P$ has 17 neutrons [1]

07.3 measure phosphoric acid into a conical flask; [1] add indicator to conical flask; [1] add ammonia solution from burette until indicator changes colour permanently; [1] stir / swirl mixture; [1] heat ammonium phosphate solution in evaporating basin until half volume remains; [1] pour into a crystallising dish and leave until crystals form [1]

08.1 high pressure favours forward reaction because there are four molecules (moles) of reactant gases [1] and two molecules of product gases **or** fewer molecules of gas in products than in reactants [1]

08.2 forward reaction is exothermic **or** reverse reaction is endothermic; [1] high temperature favours endothermic reaction or low temperature favours exothermic reaction [1]

08.3 the rate of reaction is higher at higher temperatures, [1] so ammonia is produced quickly; [1] however, yield is low, [1] so it is a compromise temperature [1]

09.1 add hydrochloric acid (or other named acid); [1] carbonate effervesces / fizzes or produces carbon dioxide gas; [1] nitrate has no reaction [1]

09.2 add dilute nitric acid and silver nitrate solution; [1] white precipitate with chloride; [1] yellow precipitate with iodide [1]

09.3 flame test; [1] calcium chloride gives red colour; [1] magnesium chloride gives no colour [1]

09.4 add sodium hydroxide solution; [1] green precipitate with iron(II) sulfate; [1] brown precipitate with iron(III) sulfate [1]

09.5 test pH of aqueous solution using named indicator or pH meter / probe; [1] ethanol solution pH 7 or appropriate neutral colour of indicator (e.g., universal indicator green); [1] ethanoic acid pH < 7 or appropriate acid colour of indicator (e.g., universal indicator red) [1]

Answers

Practical questions

01.1 thermometer [1]

01.2 **B** and **C** [1]

01.3 measuring cylinder / burette / (volumetric / graduated) pipette [1]

01.4 type of metal [1]

01.5 temperature rise / change [1]

01.6 23.2 °C [1]

01.7 **D** [1]

01.8 (type of metal is) a categoric variable [1]

01.9 (most reactive) magnesium, zinc, iron, copper [2 marks – all correct, 1 mark – 2–3 correct]

01.10 any temperature above 23.2 °C [1]

01.11 lithium is more reactive [1] so would be unsafe / dangerous [1]

01.12 **two** from: to find any anomalous results, [1] to work out a mean, [1] to check precision of results for each metal [1]

02.1 suitable scales; [1] all points plotted correctly; [2 marks – 1 mark if five points plotted correctly]
line of best fit through the points [1]

02.2 44.6 g, 50 °C [1]

02.3 line extrapolated to 70 °C, [1] 48.8 g [1] (allow value correctly read from extrapolated line)

02.4 would see a solid remaining **or** potassium chloride when added would not fully dissolve [1]

02.5 $\left(\dfrac{1\,°C}{40\,°C}\right) \times 100 = 2.5\%$ [1]

03.1 solvent line above start line [1] – dyes would dissolve into the solvent; [1] start line drawn in ink [1] – ink would run [1]

03.2 contains three different colours; [1] contains dye / colour **B**; [1] contains dye / colour **C** [1]

03.3 distance moved by dye: 1.9 cm; [1] distance moved by solvent from start line to solvent front: 2.4 cm; [1]
$R_f = \dfrac{1.9}{2.4}$ [1] $= 0.791666$ [1]
$= 0.792$ (to 3 s.f.) [1]

04.1 suitable scales; [1] all points plotted correctly; [2 marks – 1 mark if five or six points plotted correctly]
line of best fit through the points [1]

04.2 as time increases, mass lost increases; [1] then the mass lost / graph levels off [1]

04.3 carbon dioxide / a gas is produced [1]

04.4 the reading on the balance would decrease; [1] bubbles would be seen in the flask [1]

04.5 **two** from: the rate is fastest at the beginning; [1] the rate gradually slows down; [1] after 75 seconds the rate is zero or the reaction has finished [1]

04.6 measure the volume of gas produced [1] in a gas syringe / upturned measuring cylinder [1]

04.7 steeper curve initially; [1] curve levels out at 2.3 g [1]

05.1 concentration of acid [1]

05.2 0.65 (cm³) [1]

05.3 volume of acids correctly calculated as 22.25, 22.20, 22.25, 20.00; [1]
20.00 not used in calculation; [1]
$\dfrac{(22.25 + 22.20 + 22.25)}{3}$ [1] $= 22.23$; [1]
answer given to 2 d.p. [1]

05.4 would see colour change of indicator more clearly, [1] giving more accurate end point [1]

Periodic table

Key

relative atomic mass
atomic symbol
name
atomic (proton) number

Example:

1
H
hydrogen
1

1	2											3	4	5	6	7	0
																	4 **He** helium 2
7 **Li** lithium 3	9 **Be** beryllium 4											11 **B** boron 5	12 **C** carbon 6	14 **N** nitrogen 7	16 **O** oxygen 8	19 **F** fluorine 9	20 **Ne** neon 10
23 **Na** sodium 11	24 **Mg** magnesium 12											27 **Al** aluminium 13	28 **Si** silicon 14	31 **P** phosphorus 15	32 **S** sulfur 16	35.5 **Cl** chlorine 17	40 **Ar** argon 18
39 **K** potassium 19	40 **Ca** calcium 20	45 **Sc** scandium 21	48 **Ti** titanium 22	51 **V** vanadium 23	52 **Cr** chromium 24	55 **Mn** manganese 25	56 **Fe** iron 26	59 **Co** cobalt 27	59 **Ni** nickel 28	63.5 **Cu** copper 29	65 **Zn** zinc 30	70 **Ga** gallium 31	73 **Ge** germanium 32	75 **As** arsenic 33	79 **Se** selenium 34	80 **Br** bromine 35	84 **Kr** krypton 36
85 **Rb** rubidium 37	88 **Sr** strontium 38	89 **Y** yttrium 39	91 **Zr** zirconium 40	93 **Nb** niobium 41	96 **Mo** molybdenum 42	[98] **Tc** technetium 43	101 **Ru** ruthenium 44	103 **Rh** rhodium 45	106 **Pd** palladium 46	108 **Ag** silver 47	112 **Cd** cadmium 48	115 **In** indium 49	119 **Sn** tin 50	122 **Sb** antimony 51	128 **Te** tellurium 52	127 **I** iodine 53	131 **Xe** xenon 54
133 **Cs** caesium 55	137 **Ba** barium 56	139 **La*** lanthanum 57	178 **Hf** hafnium 72	181 **Ta** tantalum 73	184 **W** tungsten 74	186 **Re** rhenium 75	190 **Os** osmium 76	192 **Ir** iridium 77	195 **Pt** platinum 78	197 **Au** gold 79	201 **Hg** mercury 80	204 **Tl** thallium 81	207 **Pb** lead 82	209 **Bi** bismuth 83	[209] **Po** polonium 84	[210] **At** astatine 85	[222] **Rn** radon 86
[223] **Fr** francium 87	[226] **Ra** radium 88	[227] **Ac*** actinium 89	[261] **Rf** rutherfordium 104	[262] **Db** dubnium 105	[266] **Sg** seaborgium 106	[264] **Bh** bohrium 107	[277] **Hs** hassium 108	[268] **Mt** meitnerium 109	[271] **Ds** darmstadtium 110	[272] **Rg** roentgenium 111	[285] **Cn** copernicium 112	[286] **Uut** ununtrium 113	[289] **Fl** flerovium 114	[289] **Uup** ununpentium 115	[293] **Lv** livermorium 116	[294] **Uus** ununseptium 117	[294] **Uuo** ununoctium 118

*The lanthanides (atomic numbers 58–71) and the actinides (atomic numbers 90–103) have been omitted.

Relative atomic masses for **Cu** and **Cl** have not been rounded to the nearest whole number.

OXFORD
UNIVERSITY PRESS

Great Clarendon Street, Oxford, OX2 6DP, United Kingdom

Oxford University Press is a department of the University of Oxford.
It furthers the University's objective of excellence in research,
scholarship, and education by publishing worldwide. Oxford is a
registered trade mark of Oxford University Press in the UK and in
certain other countries

British Library Cataloguing in Publication Data
Data available

978-0-19-835941-8

10 9 8 7 6 5 4 3 2 1

Printed in Great Britain by Bell and Bain Ltd., Glasgow.

With thanks to Carl Howe for his contribution to the
Practicals support section.

Acknowledgements

COVER: PAUL D STEWART/SCIENCE PHOTO LIBRARY
p9: Olga Popova/Shutterstock; **p11**: Trevor Clifford Photography/Science
Photo Library; **p12**: ANDREW LAMBERT PHOTOGRAPHY/SCIENCE
PHOTO LIBRARY; **p22**: Ambelrip/Shutterstock; **p45**: Eco Images/Getty
Images; **p55**: MARTYN F. CHILLMAID/SCIENCE PHOTO LIBRARY; **p55**:
Praisaeng/Shutterstock; **p69**: GiphotoStock/Science Photo Library; **p75**:
Yocamon/Shutterstock; **p76**: Jaochainoi/Shutterstock; **p78**: Andrew
Lambert Photography/Science Photo Library; **p80**: Trevor Clifford
Photography/Science Photo Library; **p81**: MARTYN F. CHILLMAID/
SCIENCE PHOTO LIBRARY; **p83**: Bloomua/Shutterstock; **p83**: Robbi/
Shutterstock; **p83**: Yeko Photo Studio/Shutterstock; **p84**: David Pereiras/
Shutterstock; **p86**: Leonid Andronov/Shutterstock; **p93**: Martyn F.
Chillmaid/Science Photo Library; **p96**: Rainer Albiez/Shutterstock;
p97: Martin Kunzel/123RF; **p94**: Zacarias Pereira da Mata/Shutterstock;
p103: Worker/Shutterstock; **p105**: Alice Nerr/Shutterstock; **p109**:
Nikitabuida/Shutterstock; **p109**: Pareto/iStockphoto; **p109**: RainerPlendl/
iStockphoto; **p110**: Rtimages/Shutterstock;

Artwork by Q2A Media